KU-345-664

COAL PREPARATION

COAL PREPARATION

Proceedings of an information symposium organised by the Commission of the European Communities, held in Luxembourg on 15 November 1984

Edited by
J.K.WILKINSON
Commission of the European Communities
Directorate-General for Energy, Coal Directorate, Brussels

A.A.BALKEMA / ROTTERDAM / BOSTON / 1985

Organisation of the Information Symposium by:
The Commission of the European Communities
Directorate-General Energy
and Directorate-General for Information Market and Innovation

Secretariat of the conference and publication arrangements:
P.P. ROTONDO

The three language versions of these proceedings are being published in partnership by:
Société de l'industrie minérale, Saint-Etienne, France (French)
Verlag Glückauf, Essen, Federal Republic of Germany (German)
A.A.Balkema, Rotterdam, Netherlands / Boston, U.S.A. (English)

EUR 9681 EN
ISBN 90 6191 598 8

Published by: A.A. Balkema, P.O. Box 1675, 3000 BR Rotterdam, Netherlands
Distributed in USA & Canada by: A.A. Balkema Publishers, P.O. Box 230, Accord, MA 02018

Printed in Belgium

Contents

OPENING SESSION

- Technical activities in the solid fuel sector

TECHNICAL ACTIVITIES IN THE SOLID FUEL SECTOR

J.K. WILKINSON

Ladies and gentlemen,

You will have seen from your programme that the introductory address to this meeting was to have been given by my colleague Mr. A. De Greef, who is the Head of the Division that is responsible for questions of coal technology in the Commission of the European Communities. Unfortunately, Mr. De Greef has had health problems and has been convalescent for some time, and Mr. Reichert, the Commission's Director for Coal has, to his regret, other commitments so that it falls to me to open today's proceedings.
I should like to begin by welcoming you, on behalf of the Commission of the European Communities, to this symposium on coal preparation. The symposium is one of a long series, organized on a more or less annual basis to publicize the results of the coal research programme that is funded by the European Coal and Steel Community (ECSC). Although a wide range of research topics has been covered in the past, this is the first occasion on which such a meeting has been devoted to coal preparation. On the assumption, therefore, that many people here today will be making their first close contact with the Community's coal research programme, I thought it would be a good idea to keep to the title chosen by Mr. De Greef - "Technical activities in the solid fuel sector" - and not only to say something about that programme but also to talk about other Community activities in the field of solid fuel technology. I will also try to show where coal preparation fits into the broader picture. I have deliberately used the expression "solid fuel" here because while the ECSC programme is restricted almost entirely to hard coal, the other programmes I am going to describe fall outside the framework of the ECSC Treaty, and are not subject to that limitation. I shall speak first about the ECSC programme, since it is the longest established of the activities in question, as well as being the one that concerns us most closely today.
Under the terms of Article 55 of the Treaty establishing the European Coal and Steel Community the Commission of the European Communities has powers, originally vested in the High Authority of the ECSC, to grant funds in order to promote technical and economic research relating to the production and increased use of coal and steel and to occupational safety in the coal and steel industries. The Treaty also enjoins the Commission to organize all appropriate contacts among existing research bodies, and stipulates that the results of the research financed shall be made available to all concerned in the Community. The sort of cooperation envisaged is almost a commonplace matter nowadays but these terms of reference were something quite new when the Treaty was drafted in the early nineteen-fifties, since this was the first time that international cooperation and coordination in a specific area of research had been entrusted to a public institution. There was no question here of taking over the rôle of firms, institutes or universities in the research field. The aim was to support ongoing efforts and to make them more effective.
The history of the programme falls naturally into three parts :
- in the early years of the Community, efforts were devoted to laying the foundations for cooperation by organizing contacts between coal research

institutions in the Member States and establishing a number of international technical committees through which a wide-ranging exchange of ideas, experience and information could take place;
- the second stage began in 1959, when financial assistance for coal research projects was first granted. The research supported was initially of a rather fundamental nature, but the accent shifted gradually towards industrial applications. In the period from 1959 to 1969 the total aid disbursed amounted to 32 million ECU;
- the third phase of the programme runs from 1970 to the present time and is distinguished by three features. In the first place, the guidelines for coal research were formalized to give the programme a more definite shape. Secondly, the European Community was enlarged with the result that its coal production was approximately doubled thanks to the entry of the United Kingdom and, thirdly the energy crisis made its appearance. The last two factors resulted in the provision of considerably increased funds for Community coal research. The grand total has now reached almost 228 million ECU, 1984 included, and the annual budget is running at about 19 million ECU. These figures do not include the sums allocated for research on health protection, worker safety and industrial hygiene.

The research budget is funded from the ECSC levy on coal and steel production and it is for this reason that, as I mentioned earlier, the programme is restricted principally to hard coal. The budget is used to provide grants amounting, on average, to 60% of the costs of specific research projects. In this way the Community interest is protected and, at the same time the applicant has a financial interest sufficiently large to ensure that, in general, only promising projects are proposed and supported.
The general shape of the programme is established by a set of medium-term guidelines that is revised every five years or so. These guidelines give criteria by which the suitability of project proposals can be judged, together with a catalogue of fields and topics that are considered to be of interest. The latter is rather broadly framed, since it covers the entire spectrum of coal research from mining to utilization. In the current version, the mining engineering sector is divided into six fields :
- Development work in coal and stone;
- Methane studies, ventilation control and mine climate;
- Rock pressure and supports;
- Methods of working and techniques of coalgetting;
- Outbye operations underground;
- Modern management techniques,

while the sector of product beneficiation covers four fields;
- Mechanical coal preparation and coal transport;
- Coking of coal;
- Combustion of coal and new techniques for coal utilization;
- Coal chemistry and physics and development of processes.

Priority has generally been given to the mining sector in the allocation of funds and this is, by and large, a reflection of the Community coal industry's own priorities : as I said a moment ago, the Community's aim has been to support and coordinate ongoing efforts, rather than to take a directing rôle. On average, about 60% of the budget is allocated to mining projects. In the field of coal upgrading and utilization there has always been a strong interest in coke production because of its importance for both the coal and steel industries. Coal liquefaction and gasification have received attention, too, and there is currently an increasing interest in research on problems related to coal combustion. I will come back to these topics a little later. Coal preparation is a relative newcomer to the programme, the first project in this field being supported in 1975. Since that time, about 6,7 million ECU of aid has been allocated to coal preparation research, amounting to about 4% of the total budget, and we judged that it was now time to organize a symposium in order to make some of the results of the research supported more widely known. You will see that today's programme deals mainly with questions of the monitoring and automation of coal preparation plants, and with the problems of fines treatment. Coal mining and coal preparation lead, of course, to the production of significant quantities of waste material whose disposal and utilization give rise to serious problems for the industry. For organizational, rather than technical reasons, those problems are dealt with in a different part of the ECSC programme, and will not be discussed at this meeting.

The Commission has set up a number of committees of experts to cover the various fields of the research programme. The principal tasks of these bodies are to monitor the progress of ongoing projects and to examine new research proposals. They also provide a vehicle for the collaboration and exchange of information that is envisaged by the ECSC Treaty. You will find the names of the members of the Coal Preparation Committee listed in the programme of today's meeting, and I should like to thank them at this point for the invaluable help they have given in the planning of the symposium. The members of the experts' committees are drawn from the coal industry and its research institutions as well as, in some cases, from universities. Over and above those committees, the Commission has appointed a Coal Research Committee whose function is to advise on the choice of projects to make up each year's programme, and on the planning of new research guidelines from time to time. The members of that committee come from the coal industry - some from the production side, some from the research side - and independent experts and representatives of trade unions are also included. I mention this to show that the research programme is closely linked to the needs of the industry and therefore has a practical basis, as well as being concerned with more fundamental research. I should also add that our experience with the experts' committees has always been a particularly happy one and we are proud of the close international cooperation and friendship that has been established through these groups, which are unique of their kind.

So far I have spoken only about R & D in the ECSC framework. However, considerations of energy policy have led us into other areas of solid fuel technology. In the first place, the Commission has administered, since the end of the 1970's, a programme offering financial support to demonstration projects relating to the exploitation of alternative energy sources and to energy saving and the substitution of hydrocarbons. This scheme, which is funded from the Community's general budget, reflects a preoccupation with the idea of reducing the Community's dependence on oil, and goes beyond the stage of R & D. The aim of the programme is to show unequivocally that the technologies concerned can be exploited succesfully at the industrial and commercial level and so to promote their adoption by energy users. Because projects of the type intended are so close to the point of commercial exploitation the level of support given is lower than the 60% that I mentioned for the ECSC research programme : it is limited to a maximum of 49%, and in practice is not more than 40%. In addition, half the grant is repayable in the event of success.

Since the beginning of the scheme, support has been given to projects in the field of liquefaction and gasification of solid fuels (the expression "solid fuels" is used deliberately here because the programme is not subject to the ECSC limitation to hard coal - lignite and peat constitute indigenous energy sources that are of considerable intetrest to some of the Comu-

nity' Member States, and are therefore included in the scheme). The amount spent on projects in the liquefaction and gasification sector up to the end of 1983 is almost 100 million ECU. The emphasis of the programme has fallen on gasification. This is to be expected since, in the Community, as in other fora such as the International Energy Agency, liquefaction is seen as an expensive option, particularly in a period when oil prices have been falling. Commercial exploitation of liquefaction technology is therefore believed to be further in the future than that of gasification which, in any case, is considered to have reached a more advanced stage of development. In fact, in view of the rather long-term nature of these processes, and in recognition of the enormous development costs, the programme has recently been enlarged to include pilot-scale projects, for which the question of repayment is waived.
Since last year, the energy demonstration programme has also included projects relating to new and improved technologies for the handling, transport, combustion, treatment and storage of coal, lignite and peat and their waste products. Priority in this field has been given to combustion technology and the main is to encourage the increased use of solid fuels in the industrial and commercial sector, where the main scope for expansion is seen, rather than increased use by utilities. In 1983, 20 million ECU were committed in this area. The priority topics were fluidised bed combustion, followed by coal-liquid mixtures and the use of pulverized fuel from central preparation plants in small boilers. The Commission has recently prepared a new selection of projects to be financed from the 1984 budget and it is perhaps worth pointing out that the proposals received this year show an ever-increasing concern with environmental protection. Among the projects adopted last year, incidentally, was one related to coal preparation : it concerns the pelletization of damp fines to improve their handlability. Returning now to R&D, I mentioned earlier that some work on problems related to combustion has been supported under the ECSC programme. A series of small past and current research projects has covered the study of factors influencing emissions of materials such as hydrocarbons and heavy metals from fluidised bed boilers and more conventional industrial combustion equipment, and from power station boilers, investigation of the combustion of coal-liquid mixtures, and the application of fluidised bed combustion to some special coals. There is also a rather broad project on the improvement of coal and ash handling and the development, improvement and automation of small-to-medium scale coal-burning equipment. The general aim of all these studies is to make coal more attractive as an industrial fuel by improving the efficiency, convenience and cleanliness with which it can be handled and burned. Clearly, these aims fit closely with those of the demonstration programme that I have just described. Because of the renewal of interest in coal as a fuel for general industry it was considered desirable to give increased Community support for R & D on this subject, in addition to promoting demonstration projects. However, there were two obstacles to this : the first was the limitation of the ECSC Treaty to hard coal, a point that has been mentioned already, and secondly there are limitations to the ECSC budget resulting from the economic conditions of recent years. A way to surmount these obstacles was, as in the case of the demonstration programme, to turn to the Community's general budget - the EEC budget. In doing this, we did not act in isolation. The Community has already supported two pluriannual energy research programmes, but these were concerned with topics such as solar energy, geothermal energy, etc. and, although a small amount of work on fluidised bed combustion was supported within a sub-programme on energy saving, these two programmes were not specifically concerned with coal or other solid fuels. Some time before the second energy re-

search programme was due to expire, plans were prepared for a third one, referred to as the Non-nuclear Energy R & D Programme, and steps were taken to include in it a sub-programme on solid fuels, with emphasis on the development of new combustion technology and on environmental aspects of combustion. Approval of this new programme by the Council of Ministers has been delayed for a number of reasons including the Community's general budgetary problem but we are. hoping that matters will be resolved before long. When the EEC programme is eventually launched it will broaden the scope of Community R & D on solid fuel combustion, and will take some pressure off the ECSC coal research budget.

In preparing this address I have taken "technical activities" to refer principally to research, development and demonstration, but there is one other point that must be mentioned because it is closely linked to questions of R & D. It is, of course the problem of the environment. I am sure everyone here is a ware that the Commission has prepared a draft Directive on the limitation of emissions of pollutants into the air from large combustion plants. This is another matter that is currently being considered by the Council of Ministers. I do not want to discuss the ins and outs of this extremely complex question here : let me just say that there is clearly enormous pressure at both national and international levels to take some action on the protection of the environment. In my view it would be wrong to resist this pressure, but we should nevertheless strive to ensure that any action taken is effective, has the lowest possible cost, and is within the bounds of technical feasibility. We should also work to improve the appropriate technology as we are doing, and will continue to do in the various programmes I have mentioned. We should also bear in mind that the current draft Directive for emissions from large combustion plants is certainly not the end of the story, and that further legislation on other aspects of fuel use will be proposed before long.

At the beginning of this address I said that I would try to place coal preparation in the context of the various activities I have described. As our symposium is, after all, concerned with R & D, I shall attempt to do this by looking at the relevance of the Community's coal preparation research programme to current problems, and at the directions that such research might be expected to take in the future.

From what I have said so far, three main points emerge : the first, and most urgent, is the strong incentive to reduce emissions to the atmosphere from combustion equipment. The second is an interest in encouraging solid fuel use in industry as a means of replacing oil, and the third is the eventual introduction of solid fuel gasification and liquefaction. The first two items are of immediate interest: the last lies at a some what uncertain distance in the future.

How are emissions to be controlled? We can dismiss the problem of oxides of nitrogen very briefly here : it is a matter of flue gas treatment and careful control of combustion conditions, and has nothing to do with coal preparation. When it comes to sulphur dioxide , there are several possibilities. First, emissions can be limited by applying fluidised bed combustion with the addition of limestone or dolomite where necessary, or, secondly, by the injection of similar materials into more conventional combustion systems, although this possibility is still open to question. Thirdly, flue gas desulphurisation can be practised. This is an expensive option, and one that is probably open only to the major electricity producers. The cost will almost certainly remain too high for smaller solid fuel users, who will have to fall back on either fluidised bed combustion, which forms the priority sector of the Community's demonstration programme on solid fuel use,or on a fourth possibility, which is the use of cleaner coal. Perhaps

that is ultimately the best solution at any scale of operation, provided that such coal can be made available at reasonable cost. This is, of course, a major question for us today.
A third type of emission which I will mention briefly is that of particulate material. There is not a lot to be said here, except to mention that within the demonstration programme we are looking at the use of improved types of cyclone for the removal of fine particles from flue gases from industrial boilers, and there appears to a real possibility of making improvements at low cost in that area.
I have briefly mentioned solid fuel use in industry, and I will return to that in a moment. However, I would first like to deal quickly with the matter of gasification and liquefaction. These are techniques for the future. Their introduction depends on the answers to general questions of fuel costs and availability, but they deserve a brief mention here because it may eventually be of interest to consider the preparation of special grades of coal of particular size distribution, and perhaps enriched in the appropriate macerals, and this forms the subject of one of today's papers. At this point, I should perhaps turn to the other topics on our programme. The first of these is the automation and control of preparation plants. Developments in this field are important for two reasons : in the first place they contribute towards the improvement of working conditions and the elimination of unpleasant jobs and, in the second place, they help in the production of more consistent and better-characterised materials. This latter point will become particularly important as environmental controls become tighter.
The major item on today's programme is the problem of fines treatment. This is a problem that will not go away - it is the inevitable result of efforts to increase underground productivity. Moreover, as it becomes necessary to remove more and more sulphur, coal preparation plants will probably have to handle even greater quantities if fines, since fine crushing will probably be needed to release more mineral matter. On the other hand, coal-liquid mixtures, and particularly coal-water mixtures, are seen as a promising means of increasing coal use in industry and these in themselves demand very fine crushing, although often to a very specific size distribution. One merit of such mixtures is that they can shift the burden of drying away from the preparation plant. It is worth mentioning here, by the way, that one of the larger projects in the current demonstration programme relates to the construction of a production plant for coal-water mixtures. I should also add that there is still room for research and demonstration relating to the use of such mixtures.
With regard to the subjects that are going to be discussed today, I believe that we can look forward to steady progress, rather than to any spectacular breakthrough. Looking ahead, it seems clear that there will be a major preoccupation with coal desulphurisation, and that there is plenty of room for development in that direction. It would not be appropriate for me to go into details but perhaps, in passing, I should mention the fact that a significant proportion of the sulphur in coal is organically bound and cannot be removed by mechanical means. It is hard to imagine any form of chemical treatment that could be applied economically and in the special case of coking coal it is difficult to see how such a treatment could be applied at all without destroying the essential properties of the coal. The only answers seem to be either to gasify or liquefy the coal (but this is expensive), or to place the burden of sulphur elimination on the user, and it may well be that flue gas desulphurisation could provide the cheapest overall solution for large-scale applications.
To sum up, it appears that there will be a growing interest in desulphurisation, possibly an interest in producing special grades of coal for gasifi-

cation and liquefaction in the medium or longer term future, and certainly an increasing interest in fine coal in wet or dry form. On the other hand, the introduction of fluidised bed combustion may lead to the possibility of using coal that has undergone the minimum of preparation. It has certainly not been my intention to pre-empt what the technical experts are going to say later on today. I have simply tried to give a rather broad view. The Commission is heavily involved in solid fuel research and demonstration. While trying to take a long-term view it is obviously necessary also to keep an open mind on these matters, and today's discussions are therefore of great interest to me, as I hope they will be to the people who have come here to take part in them.

FIRST TECHNICAL SESSION

CONTROL OF COAL PREPARATION PLANTS AND COAL CRUSHING

- Measuring instruments for the continuous or rapid determination of coal properties
- Computer control of coal preparation plants
- Crushing studies for the production of coal in a form adapted to new technologies
- Morning discussion

MEASURING INSTRUMENTS FOR THE CONTINUOUS OR RAPID DETERMINATION OF COAL PROPERTIES

H. Lüdke
C. Bachmann
Bergbau-Forschung Coal Preparation Department

Summary

The control and monitoring of any coal preparation process requires sufficiently accurate information on the properties of the feed material together with an automatic means for the rapid determination of the quantities and qualities of the mass flows of feed or discharge material present in the form of loose material or pulp. An investigation was undertaken to assess the suitability of radiometric measurement, using the transmission method, for the continuous determination of the ash content of solids in pump. Americium-241 and Caesium-137 were used as radiation sources. This work, which was carried out at a semi-industrial plant, comprised the testing of the measurement system, including data collection via a process computer, investigation and quantitative evaluation of disturbance variables, temperature and pressure and finally the calibration of the measurement equipment for specific feed materials within the required limits. In the course of the investigations the continuous determination of the solids concentration, in contrast to that established by sampling, produced a residual standard deviation of approximately 1 g/l while the continuous determination of ash content achieved a residual standard deviation of about 0.4%, the measurement period being 200 s for one value. For the measurement of particle size and particle size distribution in pulp a variety of methods were tested in order to assess their suitability as on-line measuring systems. These investigations indicated that optical methods were less suited to the in-service conditions found in coal-preparation plants. The application of ultrasonic absorption for determining particle size and solids concentration in pulp proved to be much more complex than in its previous field of operation, namely the iron ore industry. Further basic research will be required before a definitive evaluation can be made. In-service tests were also conducted on a number of devices intended for measuring pulp level and for determining the solids concentration in the overflow from a wash-water thickener. One operationally-viable method proved to be that of drawing-off a side-stream from specific depths below the overflow and measuring the solids concentration with photometric pulp-measuring equipment.

1. PROCESS CONTROL USING MEASURING EQUIPMENT AT THE COAL PREPARATION STAGE

In the field of coal preparation the process-control requirements, and hence the measurement equipment specifications, can be divided into three areas:

- monitoring of individual machine groups or process stages
- control of individual machine groups or process stages
- process management with a view to optimizing the entire preparation operation on a process-related and economic basis.

Process-related monitoring and the construction of control circuits require the following parameters to be determined as part of a continuous programme:

- mass flow
- water content
- ash content
- sulphur content
- solids concentration and
- particle size distribution.

The final objective behind all efforts to introduce measurement technology is process management. One step in this direction is the use of a process computer for the rapid determination, processing and plausibility checking of measurement data. However, this requires a high degree of reliability when collecting and transmitting quantitative and qualitative values for feed, side streams and final products at the coal preparation plant. In some cases even highly-developed measuring methods do not adequately meet these requirements.

The most difficult problem is the continuous determination of qualitative characteristics, particularly with regard to ash content, water content or solids concentration and sulphur content. Since the appropriate measurement devices cannot be adopted from other branches of industry, the coal mining industry itself has been engaged in the development of such apparatus.

2. MEASUREMENT EQUIPMENT FOR THE OPERATIONAL MONITORING OF COAL FINES PREPARATION

In West Germany work began as far back as the 1960s on the use of radioactive methods for determining the ash content of bulk material. The resulting equipment has been used successfully in a number of preparation plants for monitoring and, in closed control circuits, for determining the ash content of the final product. Continuous development work is under way on measurement methods and processes which employ radioactive equipment for determining the ash content of bulk materials.

In recent years relevant measurement methods have been investigated for the determination of water content; this work has indicated that the microwave measuring system gives the least amount of deviation from measurement values which have been established by conventional means. A prototype device of this kind is to be tested under operating conditions in the near future.

The critical aspect of the research project, which was sponsored by the ECE, was the determination of the ash content and solids concentration in the flotation pulp. The following report deals with the work undertaken and results achieved when determining the particle size distribution and when testing measurement equipment for determining the solids concentration at the wash-water thickener.

2.1 Determining Ash Content and Solids Concentration in Flotation Pulp

Accurate and continuous information on the make-up of the feed material and part-products is absolutely essential for the purposes of flotation control. One way in which this data can be collected is by radiometric measurement using the transmission method.

2.1.1 Theoretical principles

The three different processes of interaction between electromagnetic, high-energy radiation and matter are the photoelectric effect, the Compton effect and pair production.

Since the atomic number of the typical ash-forming elements differ significantly from the atomic number of coal, on one hand, and that of the carrier fluid in pump (water), on the other, these interactive effects can be employed in order to determine the solids concentration and ash content in the pulp.

A suitable measurement rig can be constructed at an acceptable cost by using radioactive preparations to produce the gamma radiation. The following isotopes are particularly suitable for this:

low-reactive:	Americium 241 energy 60 keV half-life 458 years
high-reactive:	Caesium 137 energy 662 keV half-life 30.1 years.

A lower energy factor than 60 keV does admittedly bring a greater dependence on the atomic number, though it also results in a higher absorption of radiation by the pipe sections installed in the preparation plant, which in turn would produce an excessive statistical error. On the other hand, above 662 keV the level of absorption continuously falls with the result that the measuring effect, over an identical absorption path, is correspondingly less.

2.1.2 Test rig

In order to investigate the method for the rapid determination of ash content by radiometrics using the transmission system, an experimental pulp circuit was set up at the Bergbau-Forschung coal preparation research department.

A diagram of the test circuit is shown in Figure 1. A Mohno pump delivers the pulp from an agitator tank into the circuit. The radiometric measuring sections and measurement transducers for determining pressure, temperature and flow rate are installed on or in the pipe line. Two choke valves and one pressure valve serve to increase the pressure in the test circuit.

The nuclear equipment was supplied by Laboratorium Prof.

Berthold (D-7547 Wilbald). This comprised two Am-241 (300 mCi) and Cs-137 (500 mCi) emitters, two scintillation detectors and a voltage supply unit. In order to achieve a sufficiently-high level of absorption, the pipe line was taken through two right-angles. In this way it becomes possible to measure through the pipe angles in a longitudinal direction. The adsorption distances are some 20 cm in length in the case of the Americium section and 50 cm in the case of the Caesium section.

2.1.3 Disturbance parameters

When considering the adsorption equation

$$I/I_o = \exp(-\mu d)$$

several disturbance effects must be taken into account.

The equation shown above is based on a constant absorption length d. The adsorption length is fixed according to the geometric layout of the pipe. Any changes in the extension due to temperature effects can be disregarded because of the low coefficient of expansion of the materials used (steel pipe for the Caesium section and plastic for the Americium section).

Because of the construction of the test rig, and when used for in-service duties, air bubbles do in fact enter the pulp stream. Since absorption in air is minimal due to the relatively low density, there is effectively a reduction in the actual absorption length.

The density of the pulp is also a function of the temperature. While volume changes in the embodied solids as a function of temperature can of course be disregarded, temperature-dependent volume changes of the water must be considered.

Tests were conducted with pure water in order to verify, by experiment, the theoretically-assumed temperature dependence of the count rate.

The measured values and regression line for the water tests are given in Figure 2 as a function of the temperature.

The differences between the theoretically calculated values and those determined by experiment were merely 0.32%

for the Caesium measurement section and 0.61% for the Americium section. The theoretical and experimental results were thus in very close agreement.

Since water is incompressible, the pressure is not expected to affect the count rate by reason of the measurement principle. The pressure does affect the count rate, however, since the suspension being tested is always an air-solids-water mixture.

In order to investigate this effect experimentally, tests were conducted with various pressures using pure water as well as water which had been mixed with a frothing agent.

The count rates of the two measurement sections are shown in Figure 3 as a function of the particular pressure in the test system.

One can distinguish between two ranges of different pressure dependence in the two measurement sections. As the pressure rises there is a marked reduction in the count rate in the lower pressure range. Above a certain pressure the count rate assumes a linear path, as pressure increases, and is relatively low. The curve of the measurement points is identical for pure water and for water containing the frothing agent.

In the linear range the air volume no longer has an effective role to play. Here one can basically expect a constant count rate. The fact that the count rate continues to change slightly, however, can be attributed to the unavoidable mechanical stress acting on the test circuit in this pressure range (vibration and changes in geometry).

According to the results given above, it is absolutely essential to operate above the minimum pressure limit defined by the disappearance of the air-bubble effect when seeking to measure the solids concentration and ash content. The pressure-dependent fluctuations occurring in this range can be corrected by treating the pressure as a linear parameter.

2.1.4 Calibration of the measuring equipment for determining solids concentration and ash content

In accordance with fundamental tests for establishing disturbance parameters, experiments were conducted on various pulp mixtures in order to determine the relationship between the change in count rate and the solids concentration and ash content.

When calibrating the measurement of solids concentration using all available data, extremely high standard deviations occur on account of the fact that solids of differing composition were used in the various tests. This calibration process produces a standard deviation of

$$s_F = 6.17 \text{ g/l}$$

when determining the solids concentration.

However, it is physically more advisable for the tests on the materials of differing elementary compositon to be calibrated separately. In this case the standard deviations for the individual tests lie between 0.55 and 1.08 g/l.

If, with these calibrations, one examines the standard deviations for all the test series, one obtains the result

$$s_F = 0.90 \text{ g/l}.$$

By the same premise one obtains a residual standard deviation of

$$s_A = 0.39\%$$

when determining the ash content.

The results are given in Figures 4 and 5.

2.1.5 Discussion of results

As with all test procedures which are dependent on a large number of parameters, the constancy of some of these parameters is, even in the case of radiometric measurement, an important prerequisite for accuracy and consequently for the feasibility of the system. In this instance the elementary composition of the solids has a decisive effect on the accuracy of the method.

Provided that the elementary composition does not change, the solids concentration and ash content can be determined radiometrically to a relatively high degree of precision. Basically speaking, the composition of the dirt originating from an individual colliery does not change in the short term, with the result that this parameter does not usually hamper the operational viability of the radiometric measuring apparatus.

According to the principle of error propagation, the given errors represent the total error which is formed by errors in sampling and in laboratory and radiometric analysis. The sampling tests indicated that the errors produced by the conventional method represent quite a significant proportion of the total error.

If one compares the method being tested here with the universally-practised sampling technique, the advantages of continuous measurement and immediate availability of results are contrasted with the greater potential accuracy of the laboratory tests. While the new technique is not a substitute for laboratory determination, it does enable the flotation process to be monitored continuously and at relatively low cost.

2.2 Determination of Particle Size in Flotation Pulp

The rapid determination of particle size and particle size distribution is an old problem for which a number of measurement processes have been developed, mainly in the construction industry, in ore preparation for the control of grinding operations and in the manufacture of pigments, emulsions, aerosols and polishing materials.

The objective of this part of the work was to investigate commercially available measurement equipment and methods, as tried out by other branches of industry, with a view to assessing their suitability for the direct or indirect determination of particle size distribution and fines content in coal-preparation flotation pulp, where possible using on-line methods.

After a series of preliminary investigations, the tests were concentrated on a commercial measuring instrument which used laser-beam diffraction and ultrasonics to determine the particle size distribution.

2.2.1 Investigations with laser-diffraction equipment

The interaction between light and weakly-absorbent particles is dependent on the wave length used and on the particle size.

This principle is used by the Microtrac devices supplied by Leeds & Northrup. The light source is a helium-neon laser with a wave length of 0.63 μm. The following measurement ranges are offered:

* 1.9 - 176 μm
* 3.3 - 300 μm
* 31.0 - 1000 μm.

In this case the measurement ranges are each divided into 13 size categories. The Microtrac devices have only a limited measuring range, i.e. particles not falling within these ranges are only detected to a much lower degree of sensitivity and are not incorporated in the calculation of percentage ratios.

The devices are available as a laboratory version, the 'Particle size analyser', and as an operational 'on line' version, namely the 'Particle size monitor'. In both cases approximately 4 l of suspension, having a maximum solids concentration of 5 - 10% by volume depending on the fines content, is pumped continuously through a bath on to which the light beam of the laser is directed.

Flotation feed, flotation concentrate and flotation tailings were used as sample material for the Microtrac investigations since these had the particle size and distribution characteristics normally encountered in practice. The particle size distribution was established by means of wet screening and ultrasonic screening in the range $<$ 63 μm. The result is given in Figure 6.

As can be seen from the Figure, the percentage of coarser particles is, for one thing, given as too low and, for another, the percentage of particles $\leq 50\ \mu m$ in size is shown to be too small when compared with ultrasonic screening. In this case there is a problem of comparability between two measurement processes, while it is also conceivable that either additional fragmentation occurs due to the ultrasonic screening or the coagulation of the very fine particles makes itself felt during laser-beam determination.

Characterisation of the particle size and particle size distribution using a single reference value is also desirable for the control and monitoring of flotation and filtering operations. In practice the fines component (FKA) of particles $\leq 63\ \mu m$ has proved reliable. In regression calculations an attempt was made to establish a correlation between the fines component (FKA) and the values given by the Microtrac device.

The results are compiled in Figure 7. This shows the degree of certainty and the residual standard deviation for linear regressions. A satisfactory degree of correlation, with a measure of certainty of around 95% at a standard deviation of about 4%, was obtained for the dependency of the fines component (FKA) on the displayed value for the average diameter (MV) given by the Microtrac device. The Figure also illustrates the consistency of the measurements which have been taken, this being due to the sampling consistency as well as to the consistency of the Microtrac device.

2.2.2 Application of ultrasonic absorption for the determination of particle sizes and solids concentration in pulp

Ultrasonic methods have already been successfully employed to determine the fines component and solids concentration of ore pulp. The technique is based on the attenuation of ultrasonic waves as they pass through the pulp. The degree of attenuation is heavily dependent on the solids concentration and particle size of the solid particles present in the pulp and consequently can be used to measure these values. As well as the ultrasonic frequency, however, the degree of attenuation is also affected by the density of the solids. Whereas iron ore has an average density of 2.7 g/cm^3, coal has a density of between 1.3 and 2.7 g/cm^3, depending on dirt content, with the result that it is questionable from the outset as to whether or not the ultrasonic method can also be employed for coal. By reason of this, model calculations were initially undertaken with the characteristic values for coal, after which an attempt was made to confirm these calculations by means of model investigations.

It was hoped that the model calculations would illustrate the position of the viscosity and dispersion zone for flotation pulp from coal preparation plant. The first step involved the calculation of the attenuation coefficient for individual particles of density 1.2 to 2.8 g/cm^3 at an ultrasonic frequency of 0.5 MHz. The result is given in Figure 8. The coefficient of attenuation changes with the density and particle size and increases with the frequency. Above a certain particle size the effect of the solids density disappears in the dispersion zone.

Analytic tests on coal show that the above theory can also be applied here to describe the dependence between the coefficient of attenuation and the particle size.

Attempts at calibration by means of regression analysis indicate that in addition to the fines component (FKA), the entire size distribution also has an influential part to play. This seems evident since in the case of flotation products any change in particle size distribution is also accompanied by a change in the dirt content and hence in the average density of the solids.

2.3 Measurement of Solids Concentration at the Wash-Water Thickener

In order to monitor operations at the wash-water thickener it was decided to conduct tests on a suitable measurement device for determining the pulp level.

For this purpose an ultrasonic pulp-level measuring device, the USP 10 supplied by Krohne, was first installed in the centre of the rotating floor of the thickener. An ultrasonic fork, travelling up and down on a cable drum, scans the level of the thickener zone which has a specific solids concentration. The pulp-level reading did not adequately correspond with the actual height of the pulp. Furthermore, operating problems arose due to the danger of the sonic fork being broken from its mounting point.

As a replacement it was decided to install a pulp test probe on the platform itself; this device was the TAG 30/15 unit manufactured by Eur-Control. The probe was immersed 2 m below the water level. The measuring range of the device was 0 - 10 g/l solids.

Initial contamination of the probe was only partly eliminated by the subsequent inclusion of an automatic flushing system.

Process-induced faults in the current-supply system frequently put the installation out of action.

A further means for monitoring the solids level in the wash-water thickener involved the use of control lines which were positioned at depths of 1 m, 2 m, 3 m and 4 m below the overflow lip. In this case the injector principle was used to draw a quantity of pulp continuously from the appropriate depth.

Since the settling zone usually lies between 2 and 3 m, the appropriate control lines were, after the necessary conversion work, fitted with a FAK 25 continuous probe at the 3 m mark and with a TAG 30/15 immersion probe at the 2 m mark. The measurement range of the solids concentration was set to between 0 and 5 g/l.

The results obtained to date have shown that this measuring arrangement can be used for the continuous monitoring of the pulp level by setting the sampling points in the appropriate levels as well as for the regulation of the flocculant feed when this proves necessary.

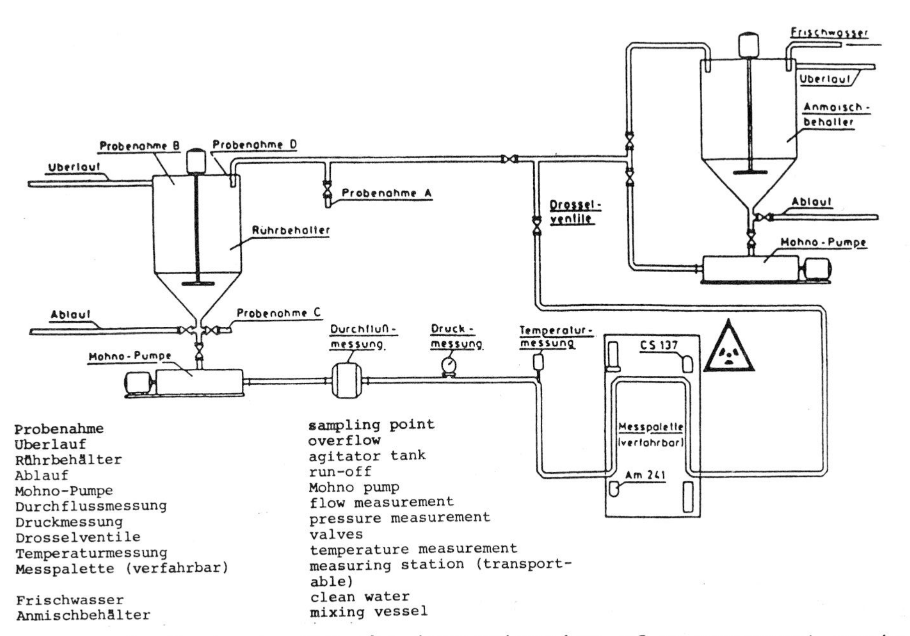

Probenahme	sampling point
Uberlauf	overflow
Rührbehälter	agitator tank
Ablauf	run-off
Mohno-Pumpe	Mohno pump
Durchflussmessung	flow measurement
Druckmessung	pressure measurement
Drosselventile	valves
Temperaturmessung	temperature measurement
Messpalette (verfahrbar)	measuring station (transportable)
Frischwasser	clean water
Anmischbehälter	mixing vessel

Figure 1. Test rig for the investigation of measurement equipment.

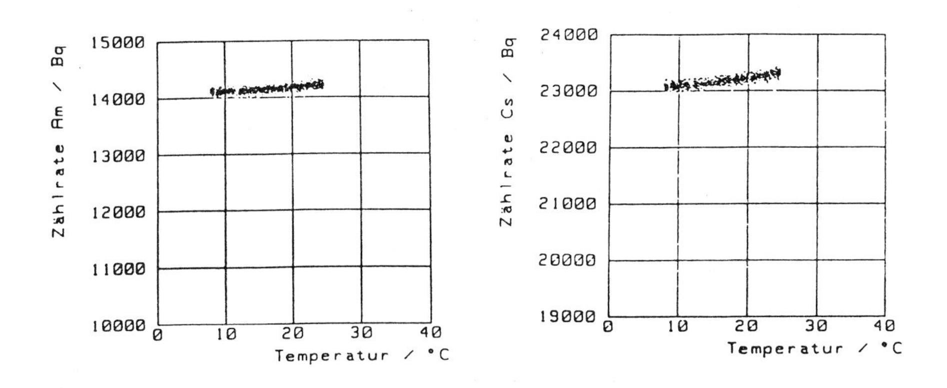

Zählrate	count rate
Temperatur	temperature

Figure 2. Temperature dependence of count rate on Americium and Caesium in water.

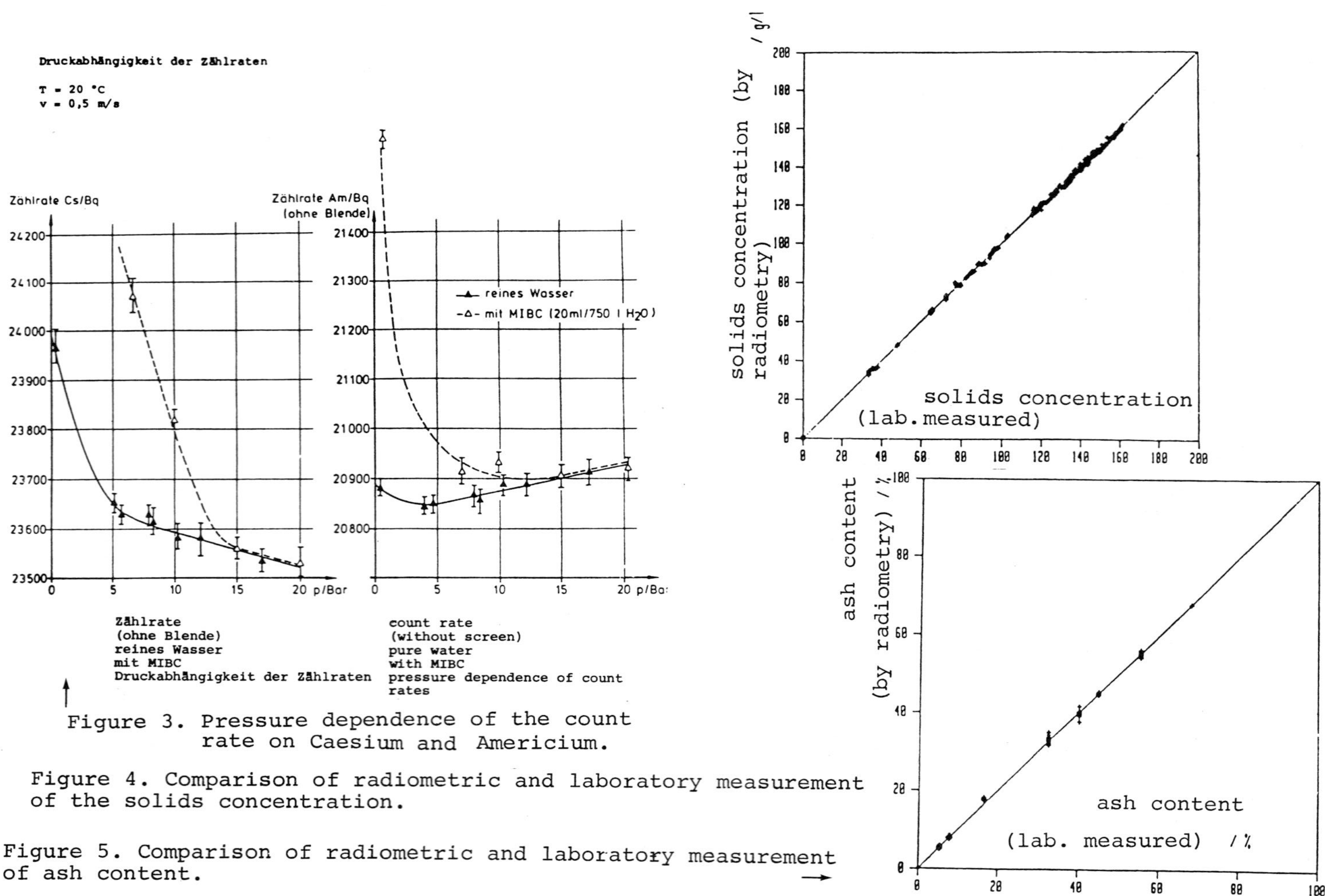

Figure 3. Pressure dependence of the count rate on Caesium and Americium.

Figure 4. Comparison of radiometric and laboratory measurement of the solids concentration.

Figure 5. Comparison of radiometric and laboratory measurement of ash content.

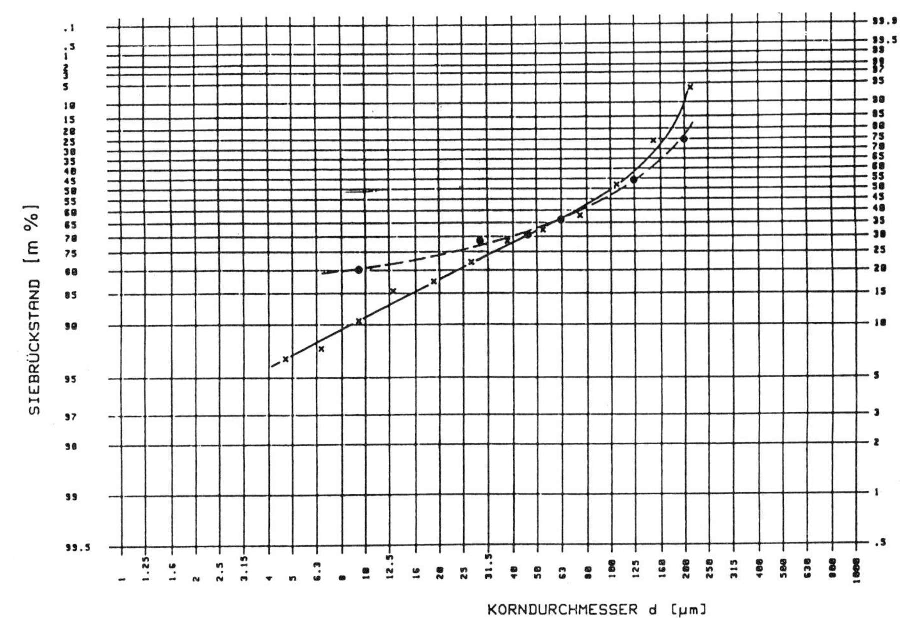

Siebrückstand screen oversize
Korndurchmesser particle diameter
Figure 6. Particle size distribution for low-volatile steam coal from Friedrich Heinrich colliery: 3.15 μm (● screen analysis, x Microtrac).

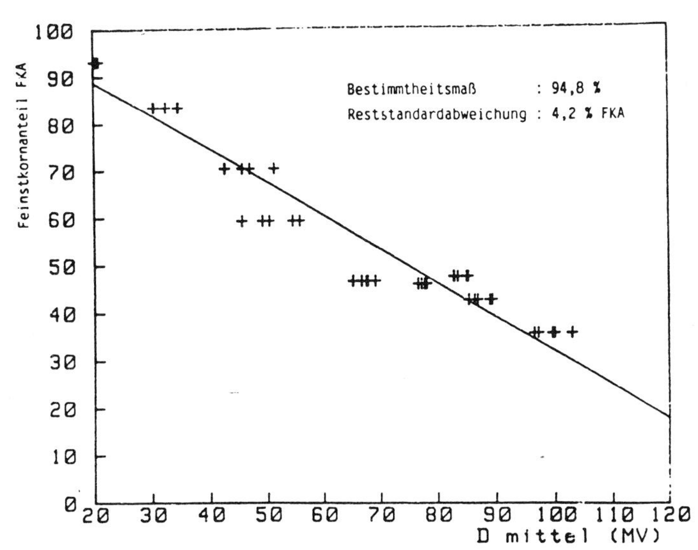

Feinstkornanteil (FKA) fines component
D mittel av. diameter
Degree of certainty: 94.8%
Residual standard deviation: 4.2% FKA
Figure 7. Correlation between fines component FKA and average diameter MV as measured by the Microtrac device (linear regression).

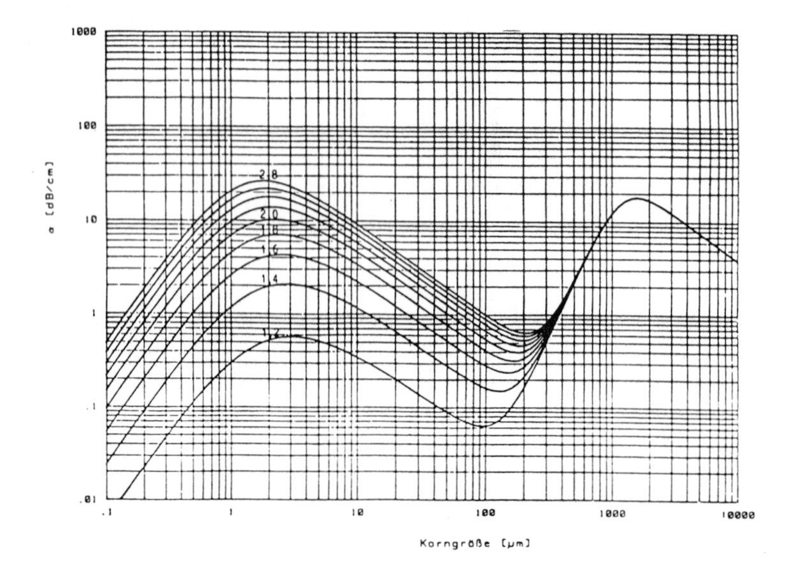

Korngrösse particle size

Figure 8. Relationship between attenuation coefficient α and particle size at 0.5 MHz for different densities (in g/cm3).

COMPUTER CONTROL OF COAL PREPARATION PLANTS

P CAMMACK, DEPUTY CHIEF COAL PREPARATION ENGINEER (R & D),
MINING RESEARCH AND DEVELOPMENT ESTABLISHMENT
NATIONAL COAL BOARD

Summary

Installations of computer controlled systems in NCB coal preparation plants have ranged from powerful minicomputers for all tasks of monitoring, control and logging to small microprocessors dedicated to precise particular duties. The experience has therefore been wide ranging, and from it a modular and distributed control philosophy has evolved which is considered best able to cater for the variety of coal preparation control requirements for large plants and small, and old and new. There are now 17 washeries in the UK equipped with computers and it has been demonstrated that with a small amount of training, operaters conditioned for many years to old hardwired systems, quickly adapt to the new control, and to use it as best suits their own particular needs.

Whilst initially the application of computers was primarily concerned with system design and implementation, resources are now devoted to exploit the potential expecially in data logging, and on-line computing facilites. It is envisaged that the biggest pay-off will result from the optimising of plant and processes, correctly ordering maintenance scheduling, and having available cost control and budgeting data on-line. All these features are a vital part of successful plant management, but which have never been possible by hardwired controls.

1. NEED FOR CHANGE

Successful automatic controls using pneumatic and electronic instrumentation, each dedicated to carry out a particular task and most likely located near to the item of plant being controlled, had been progressively introduced into NCB Coal Preparation Plants from 1960 to early 1970s. Modern washeries of that day incorporated centralised control rooms, complete with desks, for sequence group start and shut down of machinery, and where most of the data from the automatic controls and the instruments were brought together and registered by pen and chart recorders. These systems were relatively simple, readily understood, maintained and operated with fewer staff, and always associated with manual control as back up in case of controller or system failure. With few exceptions coal preparation plants in the UK were generally about 300 tph average raw coal capacity and for which these control systems were quite satisfactory.

Principally to reduce costs however coal processing facilities then tended to be concentrated in plants of larger capacity, and with a significant amount of major equipment installed in one plant, and the fact that capital costs per 100 tph had increased by almost a factor of 3/4 from 1967 to 1974, there came a greater awareness generally of the need for better utilisation of the equipment, optimisation of processes, and further reductions in manpower. Markets were also becoming more discerning requiring a greater consistency of product quality, and computer systems were considered for coal preparation plant control as a possible means of achieving these objectives. They had already been installed advantageously in a variety of other process industries including chemical and oil, and in 1975 the ECSC granted financial support to the NCB to undertake this development work for coal preparation.

2. SYSTEM DEVELOPMENT

Although the NCB was planning new washeries, it was considered prudent not to install a large computer system with its own unknown problems at a new plant and so add to the normal commissioning difficulties. Human nature would inevitably blame the computer for every malfunction of plant. Process computers in a variety of form and complexity were therefore provided at 2/3 carefully selected existing plants to provide the experience of several difference forms of approach. The installations ranged from a powerful minicomputer as at Lea Hall Colliery, covering all aspects of monitoring, display, control, logging, and reporting for a fairly large plant (600 tph), to much smaller capacity microprocessor systems as at Rawdon and Bold Collieries, principally for sequence control, but with some analogue control and limited event logging. The Lea Hall system was a Kent's K90 made up of a number of hardware components at the heart of which was a PDP11/34 computer. It provided 128 K of core memory and an additional 1.2 million words on disc store (1 K = 1,024 words, each word containing 16 bytes). In excess of 300 items of machinery provided 3,000 digital inputs and required 400 digital outputs, also 260 analogue inputs and 26 analogue outputs, and most of the data from transducers, sensors, and proving switches therefore, passed from the plant to the computer to indicate correct status before any action was initiated.

The systems for Rawdon (completed in 1977/78) and Bold (1979) were in complete contrast however, and for Bold two independent microprocessors were provided one with the maximum available 64 K bytes primarily for sequence and associated control, and the other with 48 K capacity for process monitoring and process control. There was only minimal inter-communication between the two microprocessors. Both types of system comprised colour visual display units (VDU's) for presentation of sections of the plant in flow sheet form, and alpha-numeric listings of plant status, especially alarms, Figure 1. Printers were also provided to automatically publish and record operational data.

The application of this technology to coal preparation in the UK has proceeded quickly, and to the point where there are at present a total of 17 plants controlled by computers in some form. The experience gained from the application of the various different systems, has shown that totally centralised control as adopted at Lea Hall, where all operations are managed from a single minicomputer, Figure 2, is not the best. Whilst such a system can be very powerful it is not easy to implement and is rarely understood by coal preparation plant personnel.

The concept of modular and distributed control however, which enables the process engineer to separate the different sections of plant control such as sequence, process, monitoring, display and logging, and to deal with each quite separately, offers greater attraction. Individual microprocessor systems in different parts of the plant would be dedicated to perform their own specified tasks, but would be under the control of centralised minicomputers, each to supervise a particular aspect of control as shown in Figure 3. This philosophy generally means that the software is written in a form capable of being modified by coal preparation plant personnel and without risk of corruption to the whole system of control. Because each unit is dedicated to one function a higher performance is maintained. Hardware failures in smaller distributed systems can be quickly repaired by unit replacement, and spares holding is reduced because of unit commonality. The NCB considers that a modular and distributed control system is best able to cater for the variety of coal preparation requirements for large and small plants, old and new, and has decided to use it therefore as a future standard.

3. RELIABILITY AND BACK-UP REQUIREMENTS

Whilst 100% system availability has not been obtained at the plants equipped with computers, cumulative downtime has not been a problem to cause a rethink of the strategy for control. The reliability of software is very high, and many other difficulties found in hardwired systems with relays, timers, logic circuits and so on are eliminated. Interfacing and peripheral hardware are perhaps most likely to bring a computer system down, and for this reason some form of back up is desirable. This should not be another computer however, but rather in addition to spare healthy interface cards, merely a facility to manually start and stop each item of plant from the central control room. This back-up should not be accessible to normal plant operators. As would usually be provided by coal preparation engineers, a hardwired interlock should also be provided to allow run-of-mine coal to be put to a stockpile between the colliery and the plant, in the event of complete system failure.

High plant availabilities are being achieved where there is adequate monitoring of plant equipment. At one plant where the minicomputer automatically calculates this figure, averages of total plant availability of 88% and over for 4 month periods are recorded. Excessive monitoring of non-essential signals must be resisted however, since this results in an over complex system and a reduction in plant availability through unnecessary interlock stoppages.

A clean stabilised power supply, which unfortunately is not usually available at collieries, is also of paramount importance, and whilst constant voltage transformers go part way to suppressing voltage fluctuations, they can distort the wave form to cause low voltage (5 - 25) power supply difficulties. The recommendation is to separate the colliery supply from the computer control system supply by interposing a suitable size battery and an inverter. The battery is trickle charged from the mains, thereby to act as a buffer and maintain system power during a mains failure.

4. RELATIVE COSTS

A cross-section of six new plants all with modern centralised control has been selected to examine system costs. Two had hardwired controls (A and B Table 1), and four (C-F) different levels or types of software control. Direct comparisons are not possible because features found in one system were not available in another, nevertheless it was considered valuable to extract the control system costs for each expressed as a percentage of the total system cost, also as a percentage of the plant contract price. This is shown in Table 1. An analysis of these costs show that the control system represents on average 5.5% of the plant contract sum with little differences between any of the systems examined. In the case of Colliery B the design of the control system was in fact managed in-house by the NCB, and consequently did not carry a contractors 'on costs'. These studies also highlighted that computer hardware costs have increased by less than 2% annually over the last 8 years, whereas electrical hardware has in the main carried the cost of inflation approaching 8.5%. Software systems have therefore reduced in costs relative to hardwired systems, and the indications are that with this trend continuing, ultimately they are likely to be cheaper in absolute terms. The percentage breakdown of basic control system costs also shows that cabling is significantly cheaper for the software based systems. The high design, administration and commissioning costs at Colliery F reflect a particular system complexity, but subsequently many of the techniques and sections of software developed for this plant were used again at Colliery E, which caused a considerable reduction from 42.4% to 21.5% for the same services.

5. MEASUREMENT AND MONITORING

To realise the full potential offered by any control system it is essential that transducers for coal preparation measurement and monitoring are selected with particular care and properly maintained. The alternative is the progressive failure of the control system and mistrust of instrumentation by operating personnel. In addition to the obvious requirements of satisfactory standards of accuracy, consistency and reproducability, it is necessary that instruments/transducers are chosen which will operate when used to monitor variables of difficult process materials, often located in potential damaging environments. Many of the past and present problems with transducers can be attributed to wear, corrosion or impact by the process material being monitored, but the increasing number of transducers which are successful, operate on a non-contact principal, and this allows installation away from arduous locations. Nucleonic techniques have become firmly accepted as non-contact methods for density, and with potential for level and solids flow, and concentration determinations. The application of ultrasonics is also making rapid progress as a suitable non-contact approach, also for level and flow monitoring (Figure 4). Even the successful transducers require maintenance and calibration however, but which it seems that some staff and management also, expect not to provide from the day of installation and commissioning. There is still a need therefore for education to build up confidence in the higher levels of instrumentation, and for managers and operating staff to appreciate the value and demand the information. Perhaps only then will necessary attention be given to achieve the full benefits from total computer control systems. This is a major task yet to be completed.

6. PREPARATION STAFF ATTITUDES

The introduction of modern monitoring and control systems has to a certain extent relieved staff from duties in unpleasant parts of the plant, and has generally directly provided cleaner environments for plant personnel. At one colliery for example there has been a dramatic reduction in spillage since the installation of the new control system, directly attributable to the intelligent analysis by the minicomputer. Software systems automatically process masses of information, and obviate the need for operators to keep logs of the consumption of materials and reagents, and to calculate plant availabilty and efficiency. For the older type of operator however, the intuitive sense for the operation of the plant from the noise of running items may be lost. Whilst 'noise' may be bad for health, without doubt it has been very informative. This barrier can be broken down however and at one new NCB Baum jig washery built in 1981/82 the main minicomputer control room was sited 500/600 metres away from the plant itself, and without any serious drawbacks.

The retrofitting of microprocessors to older existing plants has highlighted particularly how these forms of control have been readily accepted by plant personnel. It has been shown that the acceptance of computers to coal preparation control is not difficult even for operators conditioned for many years to old hardwired systems. With a small amount of training they quickly adapt. As one example colliery staff can now modify even to improve the system without assistance, and as further evidence at another site, the new software centralised control was completed within 14/15 months from conception through installation and commissioning, to operation by colliery staff.

7. EXPLOITATION OF COMPUTERS IN COAL PREPARATION

The benefits of installing a computer control system in financial terms, accounting for all additional extra costs, or cost reductions, benefits or disadvantages quantitively assessed, is an extremely difficult if not impossible task. Nevertheless management and plant personnel have generally accepted that software systems provide positive advantages. The flexibility of software to accommodate plant alterations quickly and cheaply is an advantage, as is the centralised clear and concise display of collected data and information by the VDU or printer. An item failed in alarm can be quickly identified and brought to the attention of the operator by message and display, with simultaneous action to prevent plant damage or spillage. Nevertheless to date the application of minicomputers to coal preparation plants has concentrated primarily on system design and implementation, and the full potential has not been appreciated or exploited. There is however an enormous facility in the processing power and memory capacity of minicomputers to improve plant performance and management by the analysis of data, integration to obtain trends, and the provision of action schedules for optimisation and plant maintenance. Similar data can also be processed to enhance budgetting controls.

7.1 Process and Plant Optimisation

Computers are used now to examine and adjust parameters associated with plant operating performance, such as the proportions of magnetite and water to constitute a dense medium suspension of the required relative density, and the correct amount of washed and untreated fines to maintain a desired blend quality. For the future however they must be used on line to determine the effect on proceeds of modifying plant conditions and changes to product qualities. The availability of products for the various markets at different plant settings can be programmed into the computers, and managers provided with all the information to allow alternatives to be considered. The potential for optimisation by computer to yield maximum proceeds from the plant whilst maintaining supplies and product qualities to customers, is therefore substantial. This has in part influenced the NCB system design strategy of front end dedicated microprocessors for control, which leaves adequate capacity within the supervisory computer for these calculations to be made. Sensitivity analysis can be more easily carried out which in the past because of the number of interacting parameters, have always been extremely time consuming and often thereby never considered. For example, in the event of a partial failure of dewatering centrifuge the effect of lowering the density of separation to reduce ash and thereby to correct calorific value for the extra moisture content, in terms of total tons and proceeds and the cost of slightly increased coal losses to reject, can all be easily and quickly assessed.

7.2 Maintenance Scheduling

While 'planned preventative maintenance' is designed to maintain maximum plant availability, it is generally based on the assumption that component or equipment wear follows a standard pattern. Unfortunately whether through faulty manufacture or other causes, breakdowns still do occur at random. With the use of monitoring transducers now available however, the store capacity of a computer is ideally suited to hold data to be used in programmes to forecast when items are likely to fail. Faulty screen springs are being detected for example to identify when the machine should be shut down to prevent more serious mechanical damage, and the wider application of condition monitoring is being pursued to give an even longer forward advance notice when particular items might need to be replaced. This technique can ultimately be developed further to check needs against the availability of spares in central or local colliery stores, and even to automatically initiate, if necessary, the purchasing authority especially where items are on national contracts. Even where suitable transducers are not available, data logging and trend analysis by computer systems enables operational events to be recorded of fault, date, time and duration, frequency patterns over various periods of time, and thereby to identify to managers priorites for maintenance. At one plant in the UK for example, the data logging system has been organised to print out at the end of each shift, day and week, the worst top six items of plant causing downtime. Whilst these reports may not provide answers or indicate decisions, they give the information to assist in more accurate decisions by the manager to achieve an overall reduction in breakdowns, improvement in plant availability, and maintenance of product quality through a continuity of operation at the optimum plant feed rate.

7.3 Cost Control and Budgetting

Cost control and budgetting are an essential part of coal preparation plant management, and although historical data is used to carry out these tasks, computers present facilities for up-to-date data to be processed even to produce forward estimates of costs (inclusive of capital and depreciation charges), for the preparation of coal in the various sections of plant. It must be beneficial to identify on a routine basis the high cost areas of preparation, whether these are caused by underloading and low plant or process utilisation, or excessive use of consumables and power charges. Obviously it will be necessary to enter certain factors such as manpower and perhaps consumable materials used manually, but computer processing power will be able to calculate on request or on a regular basis, individual values and trends to compare with budget figures. For budgetting purposes it would be possible to maintain up to date records of the capital costs of all the equipment used in coal preparation plant operations. Base date prices could be continuously modified via built in factors to correct for inflation, salary increases within the various trades and so on, and while the information may not be completely accurate, it would be readily available and sufficiently exact for budgetting control.

8. CONCLUSIONS

In the last 7 - 8 years it has been proven that the relatively new computer technology applied to other industries can also be effectively used for coal preparation plant control. From the various stages of the development work and experiences of different systems, a philosophy of system design has evolved which appears best suited to washery control.

There are 17 plants in the UK now using computers for sequence and analogue process control, data logging and recording, and the fact that the last system installed at an existing coal preparation plant was designed and to a large extent commissioned and 'de-bugged' by colliery staff, is sufficient to indicate how this technology has been accepted within the NCB.

Real benefits and pay off have not yet been identified, but it is hoped that sufficient indications have been provided to highlight where the potential is, and where therefore the NCB is concentrating continuing development effort. Inspite of the success, computers should not and will not necessarily be used on all plants, and it is firmly believed that small plants possibly less than 200/300 tph can quite adequately be controlled using hardwired systems.

Assessments will continue of the application of computer control to coal preparation plants to confirm and ensure that benefits are real in financial terms. Furthermore it is considered important to make certain that any manual tasks which remain are interesting, and that there are not new ergonomical issues caused through the 'trappings' of computers, VDU screens, printers and so on, and through no face to face contact with other staff possibly for an 8 hour shift.

9. ACKNOWLEDGEMENT

On behalf of the National Coal Board the author wishes to acknowledge the support of the ECSC in carrying out this work. Some of the prognostications are those of the author and not necessarily those of the NCB.

TABLE I - PERCENTAGE BREAKDOWN OF CONTROL SYSTEM COSTS

	COLLIERY					
	A	B	C	D	E	F
Transducers	6.6	5.8	23.3	9.0	10.4	12.5
Computer/Hard-wired control	18.3	28.9	27.8	21.2	16.1	14.5
Control Room Hardwire - desk displays, air-conditioning	10.9	4.9	12.3	7.3	8.4	5.7
Design, Administration & Commissioning	19.8	13.0	11.6	35.3	21.5	42.4
Cabling, Misc equipment, & Installation	44.4	47.4	25.0	27.2	43.6	24.9
Control System as Percentage of Plant Contract Sum	5.8	3.9	5.1	*	*	6.8

*These systems were installed on existing plants

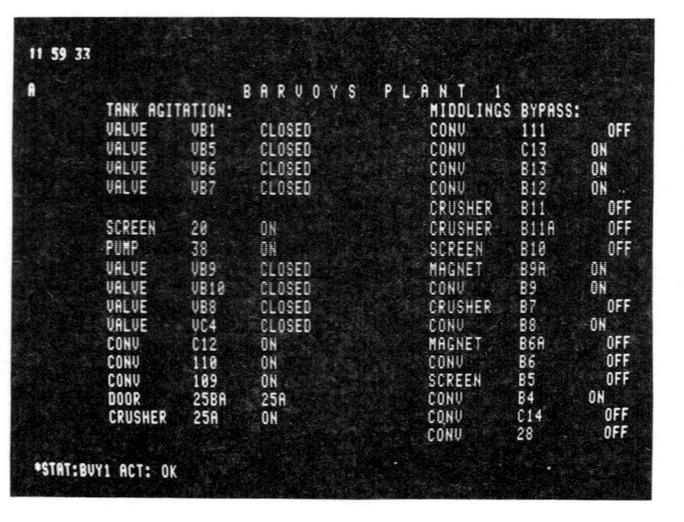

FIGURE 1. ALPHA-NUMERIC LISTING OF PLANT STATUS

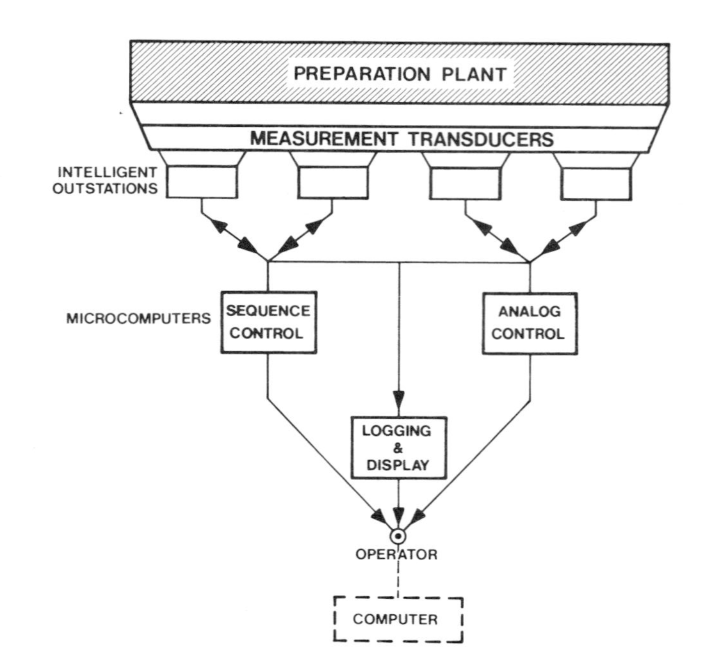

FIGURE 2. CENTRALISED CONTROL SYSTEM

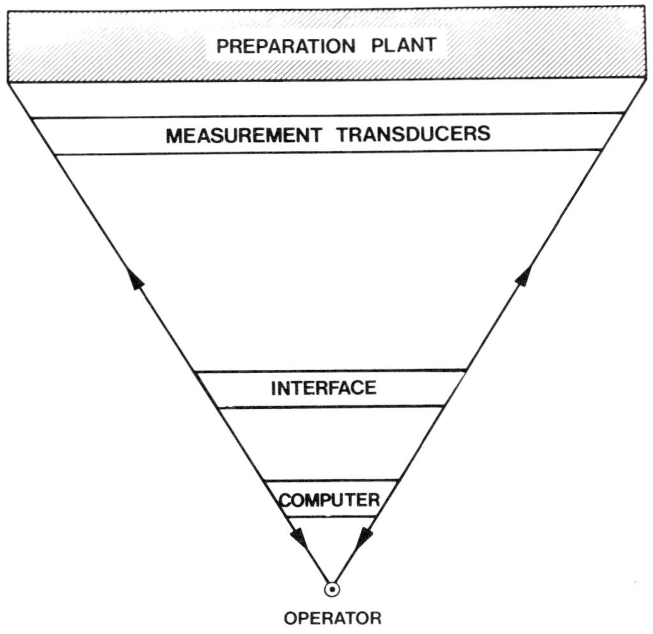

FIGURE 3. MODULAR CONTROL SYSTEM

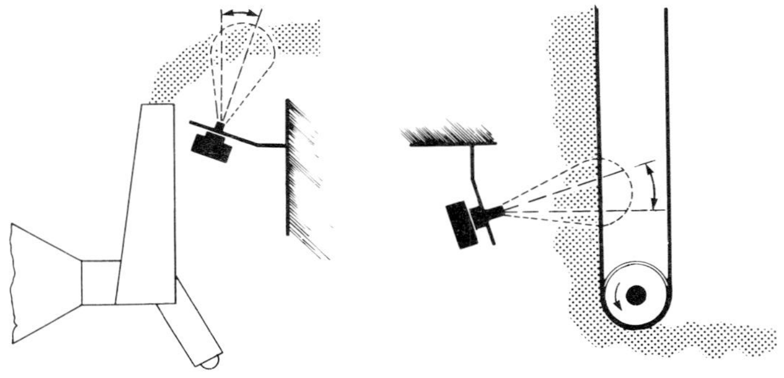

FIGURE 4. APPLICATION OF DOPPLER NON CONTACT FLOW DETECTOR

CRUSHING STUDIES FOR THE PRODUCTION OF COAL IN A FORM ADAPTED TO NEW TECHNOLOGIES

J. Leonhard
Scientific Staff Member
Bergbau-Forschung GmbH
Preparation Department

Summary

Ever since the oil crises of the Seventies the growing importance of coal as an energy raw material has been recognised throughout the world. A large number of transformation technologies to manufacture substitutes for mineral oil, natural gas and synthesis gas have been newly developed or developed further. This paper first outlines the basic process variants in gasification, combustion and liquefaction and the resultant demands made on the feed material. The selection of suitable crushers is helped by a synoptic system which is based on the criteria of applied force, size ranges of initial product and final product, degree of hardness and breaking properties of the raw material. Test results are presented for three selected machines: a swing mill, a roll mill and a dual hammer mill. Examination and consideration is given especially to mass-related crushing energies, the resultant particle sizes and surfaces for different types of coal and the pre-set machine parameters. The differing crushing behaviour is explained with the aid of a coal structure model. The mass-related crushing energy is found to be proportional to the newly created surface. The proportionality factors are dependent on the crusher used and on the type of coal.

In pressure crushing (roll mill) the behaviour does differ.

High volatile coal, fat coal and gas coal exhibit roughly the same behaviour. Surface increase is proportional to mass-related energy. For this type of coal the same proportionality factors are obtained. With anthracite the required crushing energy is only 54% and for steam coal 44% of the value needed for high volatile, fat and gas coal. This is readily explained by the coal structure model presented.

Available results show that for the many areas of coal utilisation different types of crushers have to be selected to match the properties of types of coal and indeed need to be further developed in certain areas. To be able to do this, a better understanding of the structural properties - here the strength coefficients of the coal - and of its behaviour during crushing is needed, in relation to the type and intensity of the forces to which it is being subjected.

1. INTRODUCTION

Since the oil crises of the Seventies the growing importance of coal as an energy raw material has been recognised throughout the world and its transformation into heat, steam and electricity has been intensified. Especially the new "old" processes of transforming coal to provide substitutes for mineral oil and gas are being more actively investigated in laboratory and pilot plant. Large pilot and demonstration plants are undergoing trial runs. Even though an adequate supply of mineral oil seems assured at the present time, but in future, when availability recedes and prices rise, coal will have to be ready to meet our energy supplies in a variety of forms. It will then be necessary to provide marketable and products with the help of environmentally acceptable and cost-effective transformation technologies which meet the demand for convenient, easily handled and environmentally acceptable energy forms.

2. COMPILATION OF TRANSFORMATION PROCESSES

First I would like to introduce briefly the different transformation technologies in gasification, combustion and liquefaction. Figure 1 lists the variants among coal gasification processes.

A distinction is made whether the process takes place in a fixed bed (viz. Lurgi), in a fluidised bed)viz. HTW-Winkler-process), entrained (viz. dry processes such as the Koppers-Totzek) and wet processes (Texaco).

Gasification in a iron bath is effected by blowing coal and oxygen into an iron smelter. Dosage with suitable additives will induce simultaneous de-sulphurisation.

Among the special processes is gasification with nuclear reactor heat. This is fluidised bed gasification in which the gasification heat is introduced via heat exchangers as outside heat (allothermic). Later it is proposed to feed in nuclear process heat from an HT reactor.

Combined processes are also in the trial stage, such as the GGT process, for example, in which a fluidised bed and dust gasification are integrated, and the VEW process in which pulverised coal is partly gasified and, after desulphurisation of the gas, used in a combined gas/steam turbine power station. The residual coke is utilised for secondary firing. This means fluegas desulphurisation can be dispensed with.

Among the coal liquefaction processes (Figure 2) a distinction is made between direct hydrogenation according to Bergius-Pier, extractive hydrogenation according to Pott-Broche and indirect liquefaction after gasification, followed by Fischer-Tropsch synthesis.

With direct hydrogenation finely ground coal is mashed together with oil produced in the process after addition of a catalyst and then hydrogenated once hydrogen has been added at 300 bar and 475°C. The resultant separation mainly yields liquid gas, light and medium oils. In America these projects are conducted under the name H-coal and use a higher-grade catalyst.

In Pott-Broche extractive hydrogenation the coal is treated at pressures of 100-150 bar and temperatures of 430°C with a hydrogen-shedding solvent, the so-called hydrogen-donators (for example, Tetralin). In America these projects are known as Solvent-Refined Coal (SRC) and Exxon-Donor-Solvent processes.

In the Fischer-Tropsch process synthesis gas ($CO + H_2$) from a coal gasification plant with special, selective catalysts is converted to long-chain hydrocarbons. The Sasol plants in South Africa will be familiar.

In coal combustion the processes shown in the next illustration (Figure 3) are distinguished. Grate and dust firing are well enough known as techniques. With fluidised bed combustion common variants are deep bed combustion with layers 1-2 m thick and distributor plate velocities between 1 and 3 m/s. Another variant of this technique is the expanding, circulating, fluidised bed combustion, in which the distributor plate velocities are set so high that the coal and the fluidised bed residue is overturned several times through a specially designed cyclone system. The residue acts as a heat carrier which passes its heat on to an interposed, stationary fluidised bed of the type just described. Coal-oil and coal-water suspensions are also used for firing, as a substitute for oil firing.

3. REQUIREMENTS OF FEED COAL

All these processes need a feed product which has to meet certain requirements. It is here that crushing techniques have their particular importance. The next illustration lists general crushing techniques (Figure 4). With regard to coal beneficiation processes the following objectives are important. For one, the upper limit of size distribution (for example, for pumping suspensions through pipelines and nozzles), for another, narrow size distribution (such as in fluidised bed reactors), then the specific surface (for carrying out gas-solids reactions) and the recovery of products (enrichment of maceral groups), or elimination of ash.

In the next illustration (Figure 5) these same general purposes are spelled out in concrete examples. It shows the required properties of feed coal for gasification, hydrogenation and combustion. The items on a light background indicate the requirements which have to be met by suitable crushing methods).

With fixed bed gasification (viz. Lurgi, Ruhr 100) the requirement is coal within the 40-3 mm range, avoiding fines; the proportion under 1 mm should not exceed 7%.

With fluidised bed processes the particle size should be in the 1.5-3 mm range, with fluidised fed gasification of bituminous coal with nuclear reactor heat between 0.2 and 0.5 mm. What is important here is to achieve a high proportion of equal sized particles (narrow particle band), because all particles are to have the same residence time in the fluidised bed in order to achieve a high degree of gasification, or of coal conversion.

In the entrainment process 90% of the material should be

< 90 μm and 74 μm (dry Kopper-Totzek process).

In the wet methods (Texaco) what is required is the provision of a suspension with a high solids content of around 70%. The processes operate in iron baths (Humboldt, Klöckner, Sumitumo) and require a feed < 3 mm.

For hydrogenation particle sizes in the 60-100 μm are needed, though also < 200 μm. It has been shown that what determined the requisite milling accuracy was chiefly the equipment and method and not so much reaction-kinetic considerations. An adequate stability of the coal suspension has to be assured. Coarser particles tend to sink, while valves of the high pressure pumps react sensitively to oversize. But a high proportion of very fine particles is not desirable either, because it raises the viscosity of the suspension markedly. The consequence are higher pressure losses in the pipelines and bad heat transfer in the heat exchangers. This applies to coal suspensions generally. If possible, the macerals of the exinite group should also be enriched (hydrogen saving), but minerals (pyrites excepted) and the macerals of the inertinite group (viz. fusinite), which cannot be hydrogenated and are, therefore, ballast material, should be kept low or separated out.

With power station coal, about 90% of the feed with dust firing has to be smaller than 90 μm. With fluidised bed combustion it will depend on the process variant. Deep fluidised beds work in the particle size range 1-10 μm, with a demand for a high particle uniformity here as well. Fine particles in the 20 m range serve to ignite at starting-up, but are otherwise undesirable because they are evacuated immediately without having been burnt. With circulating fluidised beds the d_{50} value of the feed product should be between 200 and 400 μm.

4. MACHINES, MATERIALS, METHODS

4.1 Selection and description of the machines used

The next illustration (Figure 6) lists the selection

criteria for the required coal milling plant. These are conditioned partly by the properties and nature of the coal to be used, partly by the requirements made of the product to be manufactured.

Which machines are suitable for the purpose? The next illustration (Figure 7) shows the possible types of crushing force depending on the raw material properties. Since coal can be described as a material of medium hardness, crushing force exerted by pressure, percussion and especially impact would all be suitable. Figure 8 presents a selection of machines appropriate for the purpose. To begin with the following mills have been investigated:

- a swing mill (Figure 9) with predominantly percussive action (swing circle 12 mm diameter / 16.6 Hz, mill pipe diameter 200 mm, length 1000 mm, feed size $\leq$ 10 mm)
- a roll mill (Figure 10) with predominantly pressure action. Measurements are taken of roller speed, pressure forces in the roller gap, the width of the gap itself and torque (drum diameter 200 mm, feed size $\leq$ 10 mm)
- a dual-rotor hammer mill with predominantly impact action (Figure 11) (190 mm wide, 490 mm impact circle diameter, peripheral hammer speed 42 m/s to 118 m/s, discharge over rod assembly classifiers or grates/screens, feed size $\leq$ 30 mm).

4.2 Designation of coals used

In the next illustration (Figure 12) the analysis data of the coals used are recorded. These were all prepared coals: Specifically an anthracite (6.67% V.M.), a steam coal (18.1% V.M.), a fat coal (24.12% V.M.), a gas coal (33.34% V.M.) and a high volatile coal (39.51% V.M.). The ash content is roughly the same between 4-6%, also the sulphur content, 0.8 - 1.2%, so that any differing behaviour during crushing has to be mainly dependent on the type of coal.

As a first means of categorising the different types of coal with regard to their behaviour during crushing these coal samples were analysed on the Hardgrove-Index (Figure 13). This yielded the following values: anthracite 39°H, steam coal 86°H, fat coal 90°H, gas coal 64°H, high volatile coal 56°H. These values have been entered on the chart. The shaded area shows the range within which German bituminous coals lie. From this it can also be seen that coals with the same V.M. content may well have differences in their grindability index, originating from the properties of the different seams, but that there may be a general trend that high rank anthracites and highly volatile gas and flame coals have greater hardness than steam and fat coals.

4.3 Measurement and evaluation methods

Granulation characteristic curves were derived from the milled products by means of a combination of wet screening and ultrasonic micro-screening. The results broadly correspond to the RRSB distribution. Parameters n and x' are obtained with the aid of a compensating line, made up of a linear regression and contiguous iteration, which takes account of the established proportions in the granular size categories and minimises error deviation. These data were used to calculate the volume-related surface S_V and the median value $x_{50,3}$.

The mass-related crushing energy (kWh/t) was calculated for the hammer mill from power consumption and throughput. For the roll mill an instrumented torque shaft was used.

5. TEST RESULTS

In the following section test results are submitted which have been obtained with these coals on the above-mentioned machines (swing mill, roll mill, hammer mill).

Particular attention is given to surface increases and the attendant energy requirements depending on the type of coal and the machine parameters.

The reason for this lies in the fact that criteria of diffusion, material conversion, heat conduction and heat transmission, so important for the heterogeneous gas-solids reactions are proportional to the specific surface. With the accuracies needed for this the grindability of coal gains its special importance because it determines the energy expended on crushing.

5.1 Swing mill

The next illustration (Figure 14) shows the results of an investigation of a swing mill (milling element: rods, granulation: 6.3-2 mm, continuous milling, product filling level: 70%). The surface increase for the coals examined is represented in relation to the feed quantity. As the feed quantity rises the newly produced surface declines. With a small feed quantity there is a marked fanning out according to types of coal. High surface values are reached with steam and fat coals, whereas anthracite produces only 1/5 of the surface increase under the same conditions.

With high feed rates such behaviour is no longer so clear. At the same time the median value of size distribution uses for example with anthracite from 65 μm at 50 kg/h to 256 μm at 150 kg/h, while it only rises from 36 to 80 μm with fat coal. Yet it is not the shorter residence time but the product filling level which is responsible for the coarsening of the milling product.

The middle granular sizes exhibit a minimum when the V.M. content is around 20%, while the specific surface S_V has its maximum.

5.2 Roll mill

Investigations of pressure crushing were carried out on a roll mill.

The next illustration (Figure 15) uses the example of high volatile coal to show the measured progression of recorded torque and pressure in the gap in relation to the roller gap and the feed quantity. As the gap becomes narrower torque and force rise exponentially. A rise in the feed rate has the same effect. With a constant feed rate there are also characteristic differences with individual coals. In the next illustration (Figure 16) the mass-related crushing energy depending on roller gap and type of coal are plotted for a feed rate of 300 kg/h. Whereas the more highly volatile fat, gas and high volatile coals display a roughly similar behaviour, the lower volatiles deviate from this. With anthracite the required crushing energy is only half as great as with these, with steam coal it is at its lowest.

Such divergent behaviour in pressure effect can be explained like this: according to van Krevelen what is characteristic for the structure of bituminous coal is the percentage of aromatic carbon and the size of the aromatic ring systems. Among low rank coal (high V.M.) bituminous coal is made up of relatively small, condensed, aromatic ring groups, which are linked by bridge structures of a non-aromatic type. The number of bridges declines with progressive coalification. At the same time the dimensions of aromatic lamellae become larger and, given some degree of extension, inter-lamellar forces begin to appear. As lamella diameters increase during coalification, graphitoid part-crystalline structures are ultimately formed.

This in turn leads to the following hardness ranges emerging, depending on the V.M. contents, which Heinze has proposed on the basis of measurements of micro-Vickers hardness.

For the $<$ 15% V.M. range he defines a so-called elastic hardness, caused by the inter-lamellar forces of the aromatic lamellae, producing a rise in the hardness curve right up to anthracite (6.0% V.M.). Because of the directional arrangement of the hexagonal stratification planes in the anthracite a distinct anisotropy is obtained here. The 15-20% V.M. range represents a transitional area (one of least hardness), in which the destruction of bridges has a hardness-reducing effect; inter-lamellar forces, however, increase hardness. In this range the term used is a brittle hardness.

In the 20-40% V.M. range hardness is largely determined by the flexible bridges and it seems appropriate to speak of a cohesive hardness, the image of a matted bridge structure giving a vivid illustration of this term.

Coals to which cohesive hardness can be attributed need a

higher energy input when subjected to pressure, while coals to which brittle hardness can be attributed (steam coal) need a lower energy input. With anthracite the anisotropic behaviour - despite greater intrinsic hardness - means that the requisite crushing energy is lower than expected.

The next illustration (Figure 17) shows this behaviour once again, but this time volatile matter is plotted against mass-related energy, depending on roller gap width and feed rate. With an average feed rate of 300 kg/h and the smallest gap width of 0.5 mm this behaviour can be seen most clearly (top unbroken curve). If the congestion in the gap eases through a reduced throughput or through a widening of the gap (the two bottom curves and the middle unbroken curve), these effects are less pronounced, as also with a very high specific load (broken curve), when these structural properties are blurred. If one plots mass-related crushing energy E_m against the average particle diameter, the so-called milling curves are obtained (Figure 18). Here, too, there is a strong fanning out according to types of coal. In order to obtain a desired fineness with more volatile coal, a considerably higher mass-related energy is needed than with low rank coals.

The surface increase rises in a linear way with the energy invested (Figure 19). Rittingers' thesis is, therefore, confirmed for the coals examined in this range. The rise of the curve depends typically on the type of coal and on the selected force. It can be seen that in pressure crushing, especially with steam and fat coal - in the middle range of volatile matter - large surface increases can be attained even with small crushing energies.

5.3 Hammer mill

Investigations of impact crushing were carried out on a dual-rotor hammer mill.

As can be seen from the next illustration (Figure 20), mass-related crushing energy rises with increasing peripheral speed of the hammers. The relative rise becomes more pronounced with rising speed with the steam, fat and gas coals, so that a curvature is obtained with increasing rise, whereas the "harder" coals such as anthracite, high-volatile coal and with middlings a linear rise is noted.

Here, also, the different coals exhibit differing behaviour (Figure 21). Anthracite has the highest energy requirement with impact crushing, steam coal the lowest.

Since the crushing behaviour of the granules depends on the physical properties, the structure of the raw material determines the milling properties. The structure can be characterised by hardness, brittleness and cohesive toughness. While with pressure crushing it is brittleness and toughness which determine the needed crushing energy, with impact and percussive crushing hardness is more responsible for the differing behaviour of coals. The basic progression is similar with different peripheral speeds of the hammers.

Higher peripheral speeds lead to greater fineness. The next illustration (Figure 22) shows the surface increase depending on the peripheral speed. The greatest increase is attained with steam coal, the smallest with anthracite. The other coals are somewhere inbetween depending on their structural firmness.

This behaviour is also expressed in the milling curves in the next illustration (Figure 23). To obtain a specified particle size with impact crushing requires four times as much crushing energy for anthracite as for steam coal. A higher working intensity, i.e. a higher peripheral hammer speed, leads to a displacement into the fines range.

If one relates mass-related energy to surface increase one obtains the energy investment as represented in the next illustration (Figure 24) as a function of the peripheral hammer speed. This means that the crushing energy is proportional to the newly created surface, regardless of how the fineness is achieved. The next illustration (Figure 25) plots surface increase against energy. For individual coals this results in a linear dependence in accordance with the Rittinger thesis.

With steam and fat coal less energy is needed for surface growth than for the harder anthracite and high-volatile coal.

With an energy input of 30 kWh/t with anthracite it is possible to obtain a surface increase of 0.5 m^2/cm^3, with high-volatile coal one of 0.8 m^2/cm^3, with fat coal one of 1.6 m^2/cm^3 and with steam coal one of 1.8 m^2/cm^3.

6. RECAPITULATION

It was shown, with the aid of the crushers investigated and the coal types processed, that crushing energy is proportional to the newly created product surface. For individual coals a linear dependence is obtained in each case, corresponding to the thesis of Rittinger. But proportionality factors are dependent on the crushers used and on the coal type. With jaw-crushing there is a fanning out of results according to the type of coal when feed rates are small, the coals hardness being established by the Hardgrove test. With coals having a V.M. content around 20% a minimum is obtained for the median value of size distribution and a maximum for the specific surface. The same applies to impact milling, where increasing energy input will achieve relatively higher surface increases for softer coals (steam and fat coals). The production of a specific particle size requires four times as much crushing energy for anthracite as compared with steam coal.

Festbett
Lurgi
Kohlegas Nordrhein(KGN)

Eisenbad
Humboldt

Wirbelschicht
Hochtemperatur-Winkler
(HTW)

Mit Kernreaktorwärme
Wasserdampfvergasung
Hydrierende Vergasung (HKV)

Flugstrom
Texaco
Saarberg-Otto
Shell-Koppers

Kombi-Verfahren
CGT Kombidruckvergasung
VEW Kohleumwandlungsverf.

direkte Hydrierung
(Bergius-Pier)

extraktive Hydrierung
(Pott-Broche)

indirekte Verflüssigung
(Fischer-Tropsch-Synthese)

1

2

Rostfeuerung

Staubfeuerung

Wirbelschichtfeuerung

Kohle-Suspension-Feuerung

3

Translations of the figures
and legends are to be found
at the end of the report.

4

5

Ziel	Anwendungsbeispiele
Breite KG-Verteilung (n (x), x)	Zuschlagstoffe Dichte Packungen
Oben begrenzte KG-Verteilung (a, b, n (x), x)	Vorzerkleinerung, Zement, Füllstoffe, Schleifmittel, Aufbereitung, Fördern aus Düsen
Unten begrenzte KG-Verteilung (n (x), x)	Flotation, Rücklauf in die Mühle Rieselfähigkeit keine Agglomeration
Enge KG-Verteilung (n (x), x)	Pigmente, Schleifmittel, Dosierung, Aufbereitung, Kohle für Vergasungsreaktoren
Bestimmte spezifische Oberfläche	Zement, Bindemittel, Kohle
Strukturänderung	Mechanochemie, Aktivierung
Bestimmte Kornform	Schleifmittel, Schotter, Füllstoffe
Aufschluß von Wertstoffen	Steine, Erde, Erze, Kohle, Kalirohsalz

① Eigenschaften	Festbett-Verfahren	Wirbelbett-Verfahren	Staub-Verfahren
Wassergehalt %	< 15 - 20	< 10 - 15	8 - 10 Br K 1 - 2 St K
Aschegehalt (wf) %	< 40	< 40	< 40
Swelling Index	< 3	< 3	-
Körnung mm	40 - 3	< 1,5 - 3 0,2 - 0,5	< 0,1 < 3
Ascheschmelzverhalten			
Sinterbeginn der Asche °C	> 900/1000	> 900/1000	-
Schwefelgehalt	keine Begrenzung		

② Eigenschaften	
Wassergehalt %	< 8 - 12 Br K < 2 - 3 St K
Aschegehalt (wf) %	< 6
Fl. Bestandteile (waf) %	> 30
Maceralgruppenanteile	
Exinitgehalt	möglichst hoch
Inertinitgehalt	möglichst niedrig
Körnung mm	0,1 - 0,06 < 0,2
Schwefelgehalt	keine Begrenzung

③ Eigenschaften	
Wassergehalt %	trocken
Aschegehalt (wf) %	7 - 25
Fl. Bestandteile (waf) %	40
Schwefelgehalt %/t SKE	0,9 - 1,5
Körnung mm	90 % < 0,090
Heizwert Kraftwerkskohle MJ/kg	22 - 30
Ascheschmelzverhalten	abhängig von der Kraftwerkstechnologie

④ Eigenschaften	
Wassergehalt %	< 20
Schwefelgehalt	keine Begrenzung
Körnung mm	< 10 - 1 WS < 1 ZWS
Heizwert J/kg	> 4200
Ascheschmelzverhalten	
Sinterbeginn der Asche °C	> 900 - 1000
Kohlenarten	keine Begrenzung

Eigenschaften und Beschaffenheit der einzusetzenden Kohlen	Anforderungen an das erzeugte Produkt
1. Stückgröße	1. günstige Restfeuchte des Produkts
2. Aschegehalt	2. je nach Verfahren optimaler bzw. tolerierbarer Aschegehalt
3. Zusammensetzung, Verwachsungen	
4. Anteil an flüchtigen Bestandteilen	3. ausreichende Feinheit
5. Wassergehalt	4. enges Körnungsband je nach Verfahren
6. Verschleißverhalten	5. Anreicherung von - für die Hydrierung - wichtigen Kompo-nenten in speziellen Kornklassen
7. Explosionsgefährlichkeit	
8. spezifische Mahlarbeit in KWh/t	

Rohstoffeigenschaft	Beanspruchungsart: Druck	Schlag	Abscherung	Reibung	Prall
hart (schleißend)	gut geeignet	gut geeignet	ungeeignet	evt geeignet	ungeeignet
mittelhart	gut geeignet	gut geeignet	ungeeignet	evt geeignet	gut geeignet
hart und mittelhart } verwachsen	ungeeignet	evt geeignet	ungeeignet	ungeeignet	gut geeignet
weich	evt geeignet	ungeeignet	gut geeignet	gut geeignet	gut geeignet
spröde	gut geeignet	gut geeignet	ungeeignet	ungeeignet	gut geeignet
elastisch	ungeeignet	ungeeignet	gut geeignet	ungeeignet	ungeeignet
zäh	ungeeignet	ungeeignet	gut geeignet	ungeeignet	gut geeignet
faserig	ungeeignet	evt geeignet	gut geeignet	gut geeignet	ungeeignet
temperaturempfindlich	ungeeignet	ungeeignet	evt geeignet	ungeeignet	gut geeignet

gut geeignet — evt geeignet — ungeeignet

6 7

Beanspruchung	Zerkleinerungsmaschinen
Druck, Schlag	Backen-, Rund-, Feinbrecher
Druck, Reibung	Walzenbrecher, Walzenmühle, Wälzmühle (Schwer-kraft-, Fliehkraft-, Fremdkraftmühle)
Schlag	Walzenbrecher (schnellaufend), Schlagbrecher
Schlag, Reibung	Rohrmühle (Kugelmühle, Stabmühle), Vibrations-mühle (Schwing-, Zentrifugalmühle)
Abscherung	Schneidmühle, gezahnte Scheibenmühle
Prall	Prallbrecher, Prallmühle ohne Mahlbahn, Luftstrahl-mühle (Spiral-, Gegenstrahlmühle)
Prall, Reibung, Schlag	Hammerbrecher, Hammermühle, Schlägermühle, Stiftmühle, Schlagkreuzmühle, Prallmühle mit Mahl-bahn, Schleudermühle (Desintegrator)
Druck, Scherung, Reibung	Rührwerkskugelmühle, Kugelmühle (langsam laufend)

8

Translations of the figures and legends are to be found at the end of the report.

9 10

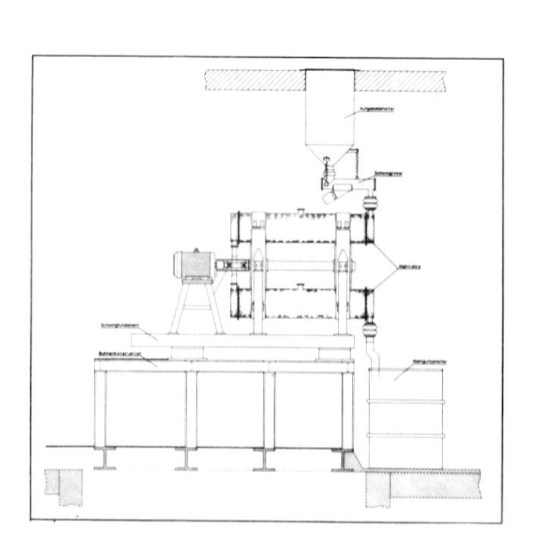

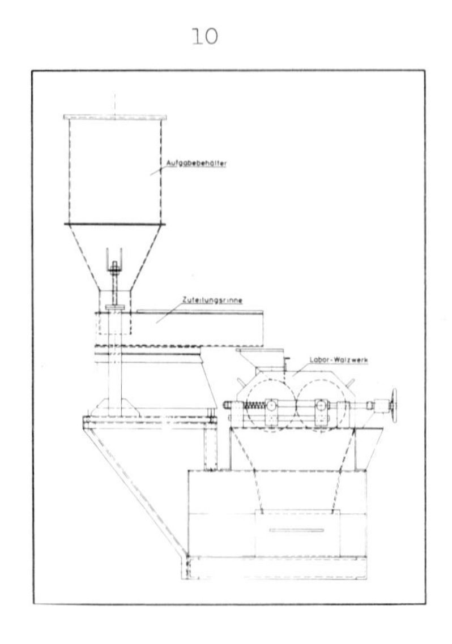

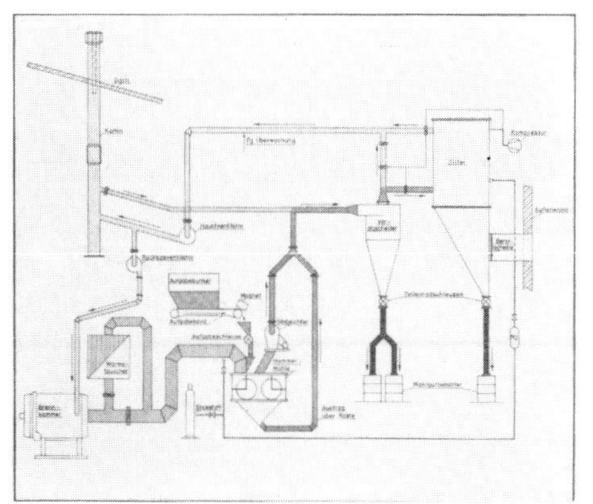

		Anthrazit	Eßkohle	Fettkohle	Gaskohle	Edelflammkohle
Kurzanalyse						
Flüchtige Bestandteile (iwaf)	%	6,67	18,14	24,12	33,84	39,51
Asche (wf)	%	5,1	3,8	3,9	3,9	4,8
Wasser	%	0,42	0,67	0,61	0,77	1,19
Elementaranalyse						
C (waf)	%	92,3	91,0	88,3	86,0	81,6
H (waf)	%	3,97	4,51	5,03	5,18	5,36
N (waf)	%	0,94	1,74	1,67	1,52	2,00
S (wf)	%	1,15	0,80	1,02	1,13	1,00
Cl (wf)	%	0,08	0,12	0,09	0,11	0,14
O(diff)(waf)	%	1,53	1,80	3,87	5,98	9,87
Petrographische Analyse						
Exinit	%	–	–	5	15	8
Vitrinit	%	64	83	80	47	73
Inertinit	%	34	16	13	36	15
Minerale	%	2	1	2	2	4

11 12

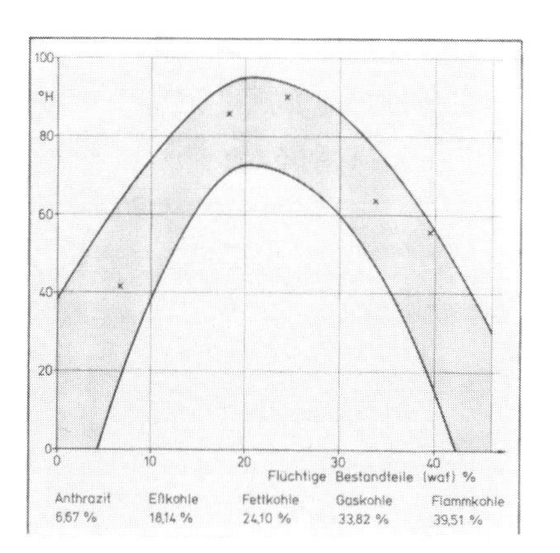

13

Translations of the figures and legends are to be found at the end of the report.

14 15

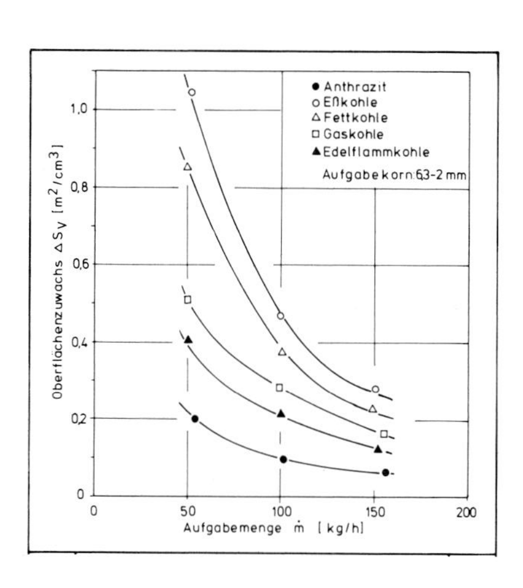

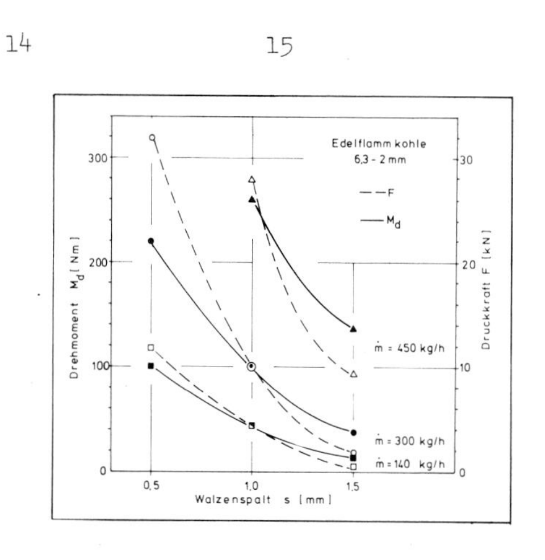

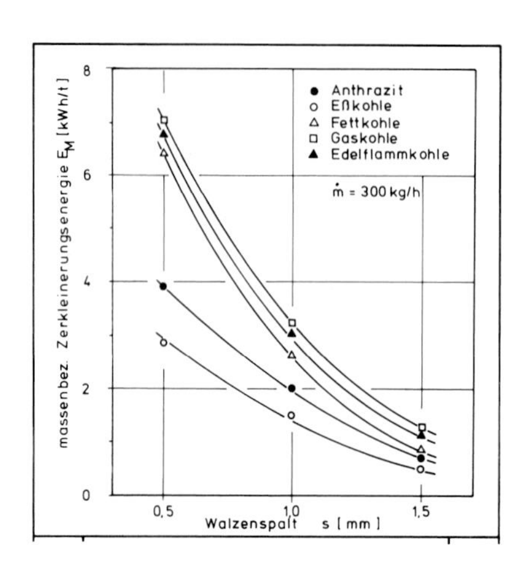

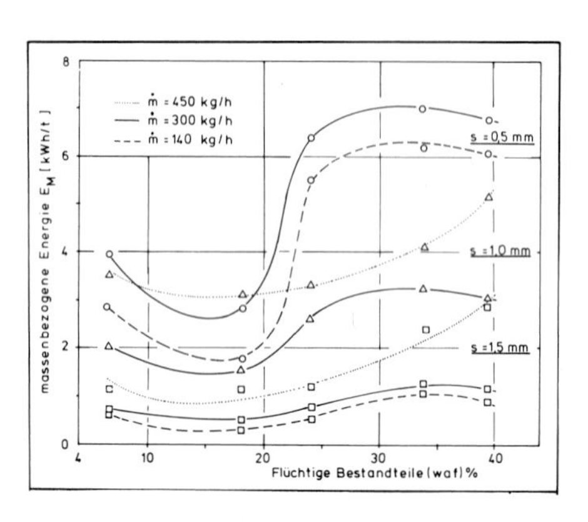

16 17

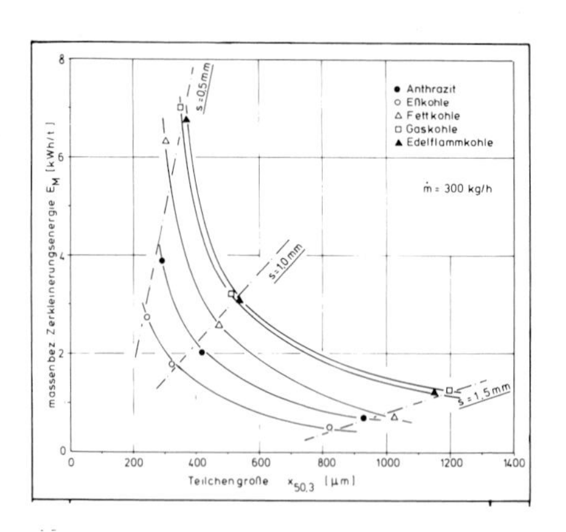

18

Translations of the figures and legends are to be found at the end of the report.

19 20

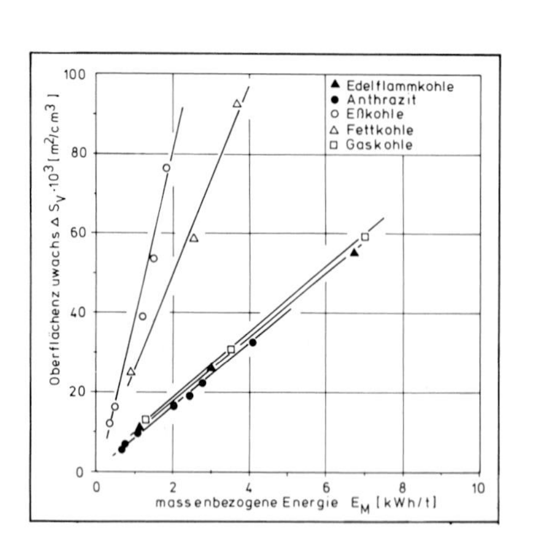

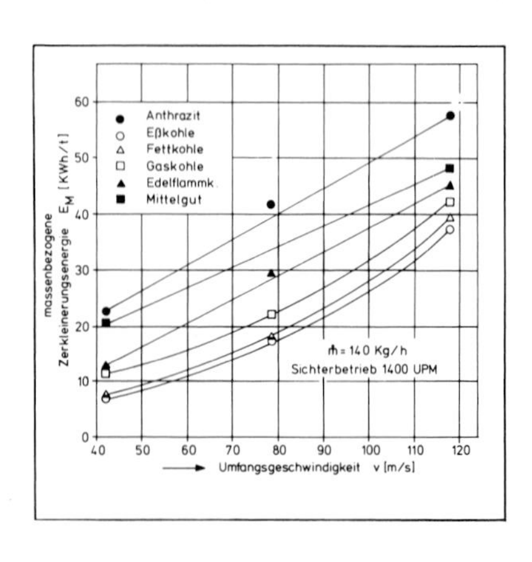

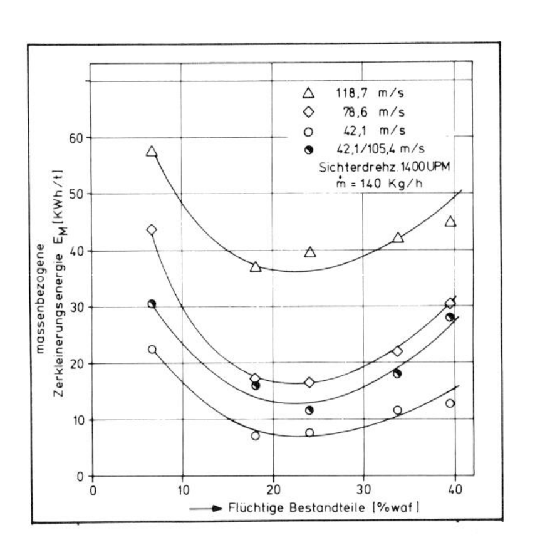

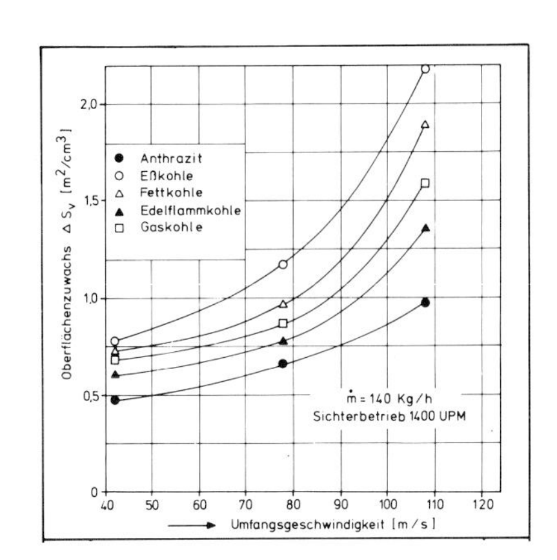

21

22

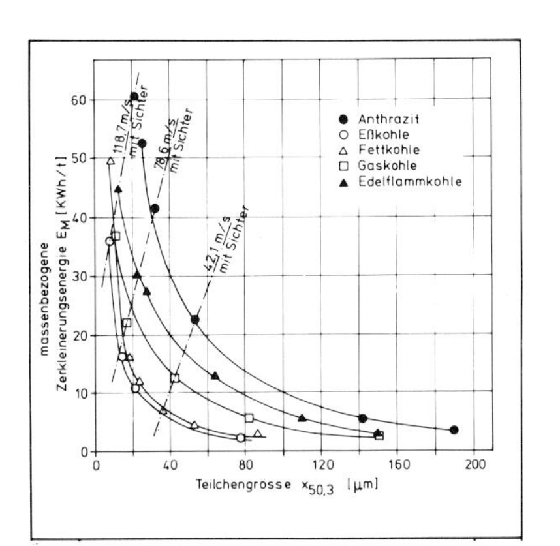

23

Translations of the figures and legends are to be found at the end of the report.

24

25

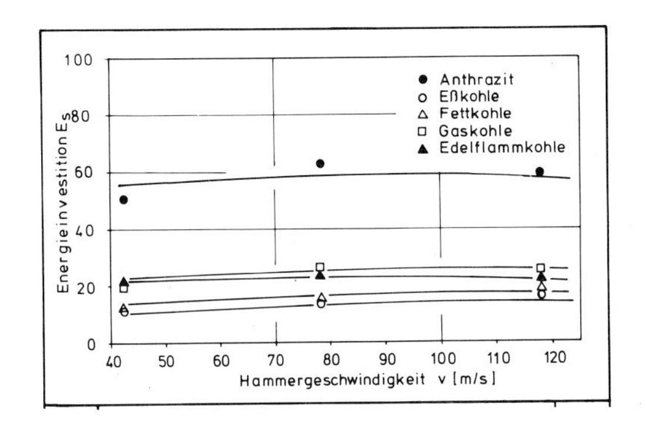

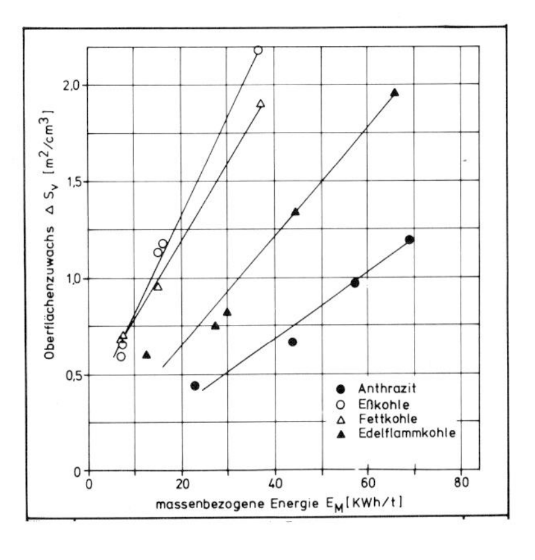

Fixed bed

Lurgi
Kohlegos Nordrhein (KGN)

Fluidised bed

HT-Winkler

Entrainment

Texaco
Saarberg-Otto
Shell-Koppers

Iron bath

Humboldt

With nuclear reactor heat

water vapour gasification
hydrogenating gasification (HKV)

Combined process

CGT combined pressure gasification
VEW coal conversion

Figure 1. Coal gasification (process variants)

direct Hydrogenation
(Bergius-Pier)

extractive Hydrogenation
(Pott-Broche)

indirect Liquefaction
(Fischer-Tropsch-Synthesis)

Figure 2. Coal liquefaction (process variants)

grate firing
dust firing
fluidised bed firing
coal-suspension firing

Figure 3. Coal combustion (process variants)

Objective	Application examples
wide size distribution	additives solid packing
upper limit distribution	pre-crushing, cement, fillers, grinding agents, preparation, nozzle distribution
lower limit distribution	flotation, return to mill, flowability, no agglomeration
narrow size distribution	pigments, grinding agents, dosage, preparation, coal for gasification reactors
specific surface	cement, binders, coal
structure change	mechano-chemistry, activation
specific size	grinding agent, ballast, fillers
recovery of valuable products	stones, earths, ores, coal, crude potassium

Figure 4. Crushing techniques and their objectives

1 Properties			2 Properties		
	Fixed bed	f.b. dust			
water content	%		water content	%	
ash content	%		ash content (wf)	%	
swelling index			V.M. (waf)	%	
granulation	mm		macerals		
ash fusibility			exinite		as high as poss.
ash sinter start	°C		inertinite		as low as poss.
sulphur content			granulation	mm	
			sulphur content		no limitation

3 Properties			4 Properties		
water content	%	dry	water content	%	
ash content (wf)	%		sulphur content		no limitation
V.M. (waf)	%		granulation	mm	
Sulphur content	% tce		calorific value	J/kg	
granulation	mm		ash fusibility		
calorific value	Mj/kg		ash sinter start	°C	
power station coal			coal types		no limitation
ash fusibility	depends on technology				

Figure 5. Properties of feed coal for
1 gasification 2 hydrogenation
3 electricity generation 4 fluidised bed combustion

Properties and character of feed coal	Requirements of final product
1. Lump size	1. favourable residual moisture of product
2. Ash content	2. optimal or tolerable ash content, depending on process
3. Composition, intergrowth	3. Adequate fineness
4. Share of V.M.	4. Narrow size range, depending on process
5. Water content	5. Enrichment of components important for hydrogenation in special size categories
6. Wear effects	
7. Explosion danger	
8. Specific grinding in kWh/t	

Figure 6. Selection criteria for coal grinding plant

Raw material property	Pressure	Percussion	Shearing	Friction	Impact
	Type of force applied				
hard (wearing)					
medium hard					
hard & medium) intergrown					
soft					
brittle					
elastic					
tough					
fibrous					
temperature-sensitive					

well suited   possibly suitable   unsuitable

Figure 7. Type of force applied depending on raw material properties.

Force	Crushers
Pressure, percussion	jaw, rotary, fine crushers
Pressure, friction	roll crushers, roll mill, rolling mill (gravity, centrifugal, external force)
Percussion	roll crushers (high speed), percussion crushers
Percussion, friction	tube mill (ball mill, rod mill), vibro-mill (swing, centrifugal mill)
Shearing	impeller mill, toothed disk mill
Impact	impact breaker, impact pulverizer without grinding track, air jet mill (spiral, reflected jet mill)
Impact, friction, percussion	hammer crusher, hammer mill, pugmill, pinned disk mill, beater mill, impact pulverizer with grinding track, disintegrator
Pressure, shearing, friction	agitator ball mill, ball mill (slow speed)

Figure 8. Modus operandi of crushers

Figure 9. Test installation for crushing hard coal - swing mill

Figure 10. Test installation for pressure crushing of hard coal

Figure 11. Process flow-sheet of grinding dryer for hard coal with self-inertisation

	anthracite	steam coal	fat coal	gas coal	high volatile coal
Instant analysis					
V.M. (waf)					
Ash (wf)					
Water					
Ultimate analysis					
C (waf)					
H (waf)					
N (waf)					
S (wf)					
C (wf)					
O (diff)(waf)					
Petrographic analysis					
Exinite					
Vitrinite					
Inertinite					
Minerals					

Figure 12. Analysis data of the coals investigated

V.M. (waf) %

anthracite steam coal fat coal gas coal high vol. coal

Figure 13. Grindability of bituminous coals (Hardgrove)

anthracite
steam coal
fat coal
gas coal
high volatile coal

feed 6.3-2 mm

Figure 14. Surface increase S_V as a function of feed quantity (swing mill)

high volatile coal
6.3 - 2 mm

Figure 15. Torque and applied pressure depending on roller gap and feed quantity

Figure 16. Mass-related crushing energy depending on roller gap and type of coal (for symbols see Figure 14)

Figure 17. Mass-related crushing energy depending on type of coal, gap width and feed rate

Figure 18. Mass-related crushing energy E_m as a function of particle size $x_{50,3}$ = median value (roll mill). (For symbols see Figure 14)

Figure 19. Surface increase ΔS_V as a function of mass-related energy E_m (roll mill). (For symbols see Figure 14)

Figure 20. Mass-related crushing energy E_m depending on peripheral speed and type of coal (hammer mill). (For symbols see Figure 14)

Figure 21. Mass-related crushing energy of different coal types with different peripheral speeds (hammer mill)

Figure 22. Volume-related surface increase depending on peripheral speed (hammer mill). (For symbols see Figure 14)

Figure 23. Mass-related crushing energy as a function of particle size $x_{50,3}$ = median value (hammer mill). (For symbols see Figure 14)

Figure 24. Energy investment depending on hammer speed v. (For symbols see Figure 14)

Figure 25. Surface increase ΔS_V as a function of mass-related energy E_m (hammer mill). (For symbols see Figure 14)

Morning discussion (by J.K. WILKINSON)

Mr. MANACKERMAN (National Coal Board) queried the figures on coal size for deep fluidized bed combustion cited in Mr.LEONHARD's paper. Mr. LEONHARD explained that the paper contained a misprint and that the correct size range was 1 to 10 mm.

In answer to a question from Professor SMIRNOW (Westfälische Berggewerkschaftskasse) on the accuracy with which the energy consumption in crushing had been measured, Mr. LEONHARD stated that this parameter had been measured in terms of the difference between energy consumption during idling and crushing. The figures were reasonably reliable, although the measurement was easier to make for roll crushers than for hammer mills.

Mr. BOCK (Universität-GHS-Essen) drew attention to Mr. LEONHARD's diagram relating crushing energy to the coal's volatile matter content. This showed that higher volatile coals were more easily crushed, although other work indicated that crushing was easier for coking coal. Mr. LEONHARD replied that the diagram showed crushability in terms of the Hardgrove index, which was subject to broad variation, so that there could be an overlap between coals.

Mr. REUTER (Chemische Fabrik Stockhausen GmbH) asked whether the discrepancies reported by Mr. LUEDKE between the size distribution of fines measured by sieving and by the Microtrac technique could be attributed, in part, to microflocculation caused by electrical effects. Mr. LUEDKE said that this could be the case, but that the measurements had not been carried out in sufficient detail to reveal such phenomena.

Dr. WILCZYNSKI (Bergbau AG Lippe) asked Mr. CAMMACK whether the National Coal Board had applied computer control to the dosage of additives. The latter replied that, while individual control systems were used for dosing flocculant and magnetite, these operations were not yet under computer control except in a single case where a programmable logic controller was used to regulate magnetite make-up.

Mr. ROESNER (Exploration und Bergbau GmbH) asked what criteria were used when optimization calculations were carried out as a part of the computer control of coal preparation plants. Mr. CAMMACK answered that such calculations were, at present, performed off-line in most countries. In the case of the

National Coal Board, the data used were derived from extensive performance tests on individual plant items (jigs, dense-medium baths, etc.) so that the parameters used to predict performance included all the practical variations and errors that occurred in plant operation, and were not simply theoretical factors. Mr. CAMMACK believed that such calculations could be carried out on-line in the future, since they could give a rapid prediction of the effect on product quality of variations in process parameters such as, for example the density of dense media. Mr. ROESNER asked whether the question of costs and profits entered into the calculations that Mr. CAMMACK had described. Mr. CAMMACK replied that such considerations should be included although this had not yet been done. One of the potential benefits from coal preparation plant control by computer, which could not be obtained from hard-wired systems, was that with adequate computer capacity it should also be possible to include cost factors in order to maximise proceeds. This was an aim for the future. Mr. JENKINSON added that, in designing preparation plants, the National Coal Board achieved optimization through computer simulation of alternatives. The values of products were used in the design procedure to maximize yields and proceeds. The future aim was to store all the relevant information and to use it on-line. Blending of the 76 million t/year of coal delivered to power stations in the UK was already optimized. For each mine concerned an optimum point existed,at which proceeds were a maximum, and which was reached by automatic blending systems operating within set limits on ash content and calorific value. One of the papers to be presented at the symposium dealt with the improvement of the continuous measurement of product ash content in order to improve the accuracy of such systems. However, although better accuracy was desirable, optimization could still be achieved with the comparatively rough-and-ready means available.

SECOND TECHNICAL SESSION

FINES TREATMENT

- Study of means of improving the wet sizing of coal fines
- The development and application of the rotating probability screen
- Recent developments in the dewatering of fine and super-fine coal products
- Membrane pressure plate filtration
- The influence of surface-active agents in washing water on flotation and flocculation
- Afternoon discussion
- Closing address

STUDY OF MEANS OF IMPROVING THE WET SIZING OF COAL FINES

J.L. DANIEL

Engineer in the Preparation-Handling Division
CERCHAR

Summary

The experiments carried out within the project of possible improvement in underwater sizing of fine coal comprised two stages of work.

The first stage was devoted to assessing the sizing equipment already in service in the industry. The machines selected as examples were of two different types:

- a flat screening panel fitted with a cloth deck,
- a curved screen (or curved grate) fitted with a slotted screen plate.

The screening panel provides fairly accurate cuts typified by an average probable deviation mostly between 0.04 and 0.05. Effective separation area is close to 2/3 of the mesh aperatures in the deck. These results are applicable to a slurry of average clay content with a specific throughput of around 25-30 m^3/h per m^2 of screening surface.

The curved grate produces rather less accurate separation than the screening panel with an effective separation area somewhere between half and two thirds of the bar spacing.

The specific throughput of slurry is appreciably higher than that of the flat panel: 45 m^3/h per m^2.

In the second stage of this research work it was planned to examine possible ways of improving sizing by achieving greater accuracy in separation and by reducing the disadvantages caused by fouling of the screening surfaces.

Several methods have been tried but most of them did not produce the hoped for effect. Only two items of equipment proved promising:

- a support arrangement for the screening panel deck which, by reducing capillarity effects, eases the passage of water over the screening surface,
- a vibrator which simply uses the water entraining the undersize as the motive force. The vibrations produced in this way keep the screening surface free of all clogging.

1. INTRODUCTION

Posing the problem - Object of research

Enforcement of environmental regulations and the constraints imposed by technical advances oblige coal users to become more and more stringent in their requirements to the quality of products supplied to them.

Processes for treating coarse fractions have achieved a degree of accuracy which it is difficult to improve upon.

On the other hand, progress needs still to be made in the field of treating coal fires of which a marked increase has been noted in the unwashed coal as a result of the intensive mechanisation of winning operations.

The methods used to bring about the separation of coal and dirt - jig, cyclone, table - provide a more accurate cut-point the more compact the range of size in the feed is and, as a consequence, the more accurate the sizing.

Conventional hydraulic classifiers can respond to this requirement only inadequately because the size separation is here effected by equal falling.

Screening machines with fixed flat or curved grates yield more accurate separation, but their proper functioning is often disrupted by various incidents, in particular by clogging when treating certain products - coal with a high clay content especially.

For certain new technologies problems of sizing also arise prior to utilisation.

For example, the manufacture of coal-water mixtures usually involves a wet crushing operation calling for exercising size control over the crushed product with the elimination of particles larger than a given mesh size.

The research work described below aimed to improve the quality of size separation of fine fractions before their subsequent cleaning (or utilisation).

Experiments were carried out in two areas:

1) Assessment of existing equipment;
2) Tests to improve separation accuracy and a search for means of reducing clogging effects.

2. ASSESSMENT OF SIZING MACHINES

The equipment used as examples were chosen to represent different types:

- a flat screening panel fitted with a cloth deck,
- a curved screen (or curved grate) fitted with a slotted screen deck.

2.1 Flat screening panel

BRGM patent - machine supplied by SAULAS-PARIS

2.1.1 Description

The screening panel basically comprises:

- a supple cloth made of synthetic fabric stretched over a stainless steel supporting plate with 5 mm square meshes. These two surfaces - cloth and deck - can be tilted between 30 and 45° (from the horizontal),
- a supporting underframe fitted with deck tensioners.

The slurry to be screened is distributed across the entire width of the deck by means of a feed box mounted at the upper end of the underframe.

The fine particles are entrained by the bulk of the liquid and pass across the mesh of the screening surface. The coarser particles build up into a slope which slides on the deck, pulled along by its own weight. Separation is brought about by gravity alone without the help of any auxiliary energy. Figure 1 illustrates the above data.

2.1.2 Efficiency

The screening panel provides rather accurate cuts typified by an average probable deviation usually within the 0.04 and 0.05 range. The bar chart of the frequency of probable deviation obtained is reproduced in Figure 2. Figure 3 gives an example of a recorded separation curve. This graph confirms the absense of "oversizes" in the screened product, i.e. particles of size larger than the mesh dimensions of the deck. This characteristic is greatly appreciated in certain user applications.

The effective separation area is close to 2/3 of the mesh apertures of the deck.

These results were obtained with a slurry of medium clay content, fed at a specific throughput rate of 25 to 30 m^3/h per m^2 of screening surface.

The parameters capable of affecting separation accuracy include the following:

- the inclination of the screen, the optimal value of which is around 40° for the coal in question,
- the percentage of fine particles under 50 microns. Separation accuracy decreases once this percentage goes over 25%,
- the feed concentration. The overall effectiveness of sizing decreases once concentration exceeds 400 g/l.

On the other hand, the mesh shape - be it square or rectangular - does not appreciably affect either the functioning of the screening panel or the efficiency of separation.

2.2 Curved grate

2.2.1 Description

The curved screen tried out is a slotted screen plate in the form of a cylindrical segment on a horizontal axis. The open angle of the grate arc equals 60°.

The bars are trapezoidal in cross-section and form the generators of the cylinder, being fitted at right angles to the flow direction as can be seen from Figure 4.

Two grates were tested in turn:

- grate No. 1 with a bar spacing of 0.3 mm,
- grate No. 2 with a bar spacing of 0.15 mm.

2.2.2 Efficiency

The curved grate yields slightly less accurate separation

than the screening panel.

The probable deviation is usually between 0.05 and 0.08, the mean value being 0.06.

On the other hand, the specific throughput per m^2 of screening surface is distinctly higher, of the order of 45 m^3/h per m^2.

Effective separation area lies between half and 2/3 of bar spacing.

Any thickening of the slurry (beyond 300 g/l) leads to reduced separation accuracy.

Tests have also shown that the curved grate is more effective at trapping coarse product when this is present in small proportions than in eliminating a small quantity of fine particles dispersed in a slurry made up of coarse particles.

The percentage of marginal particles did not vary sufficiently during our tests to enable a clear trend to be deduced. In any event it is very likely that effectiveness diminishes when the percentage of marginal particles increases.

3. TESTS TO IMPROVE SEPARATION ACCURACY - COMBATING THE CLOGGING OF SCREENING SURFACES

Product screening on fixed screening panels or curved grates makes it possible to achieve much more accurate cut-points than those obtained by hydraulic classifiers such as cyclones. But because of the inherent principle of separation by means of screening, difficulties abound as soon as fine and sticky product has to traverse the mesh or the interstices of small dimension.

To counteract these drawbacks a number of devices have been tested.

3.1 Screening panel

3.1.1 Spraying by mobile distributor

The screening surface is swept by water jets provided with a travelling motion.

Cut-point accuracy is improved - Epm = 0.043 to 0.049 - as is the quality of the overflow which retains less displaced fine material.

3.1.2 Screening deck resting on a porous bed

This porous bed consists of plastic balls and unclogging of the deck was to be effected by repeated shocks administered by the balls as they were raised by the level of the liquid as it flowed.

Technical difficulties prevented this device being implemented.

3.1.3 Device with suction action (see Figure 5)

Here the screening surface rests on a grooved plate which serves to evacuate the filtered water. The suction action created by this flow was to reduce the effects of capillarity.

Promising results were noted both with regard to separation accuracy (average probable deviation fluctuating between 0.025 and 0.032) and keeping the screening deck in proper condition.

3.1.4 Vibrating drum

Figure 6 gives a diagrammatic view of the device.

The screening deck is activated by vibrations administered by a vaned drum rotated by the water entraining the undersize.

The vibrations keep the screening surface clear of all clogging and give a slight improvement to separation accuracy; the improvement is seen both in the average probable deviation (0.017 to 0.132) and in sizing efficiency (91.3 to 92.1).

3.2 Curved gate

Three devices were tested.

3.2.1 Vibration of shell

An electromagnetic vibrator was installed on the shell.

3.2.2 Arrestor paddle for overflow

A paddle which is freely rotatable around a horizontal

axis and balanced by a counterweight was fitted in the "overflow" compartment to check the flow of this product by maintaining a permanent slope of coarse products in the lower part of the grate.

3.2.3 Shocks to grate from pneumatic jack*

A piston controlled by a pneumatic jack hits the grate at intervals which can be regulated from 10 to 60 blows per minute.

Of these three devices only the pneumatic jack held out some promise in avoiding clogging and ensuring better water passage across the interstices of the grate.

4. CONCLUSION

The research work carried out within this project made it possible to characterise the aptitude for screening fine coal by two types of equipment: screening panel - curved grate.

The screening panel provides more accurate size separation but its specific throughput is lower than that of the curved grate, though the latter is less accurate in sizing particles.

Both items of equipment are liable to clogging by clayey products. Several devices designed to counteract this drawback were tested. Two of these have proved to be effective:

- the device with suction action,
- the vibrating drum,

and should enjoy industrial development because of the simplicity of introduction.

* RAPIFINE screen supplied by DORR OLIVER - PARIS LA DEFENSE.

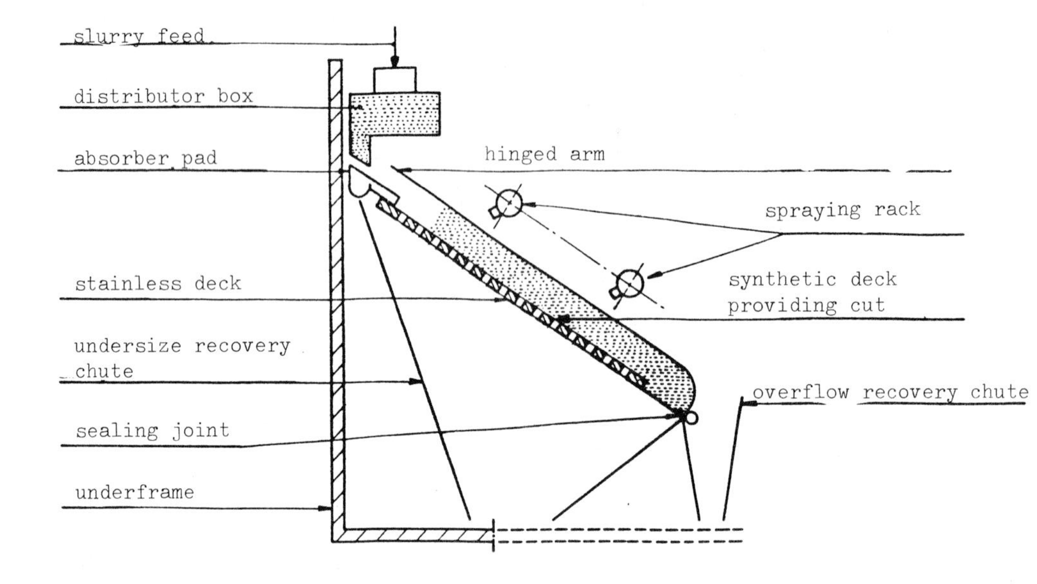

Fig. 1 Screening Panel - Diagrammatic view.

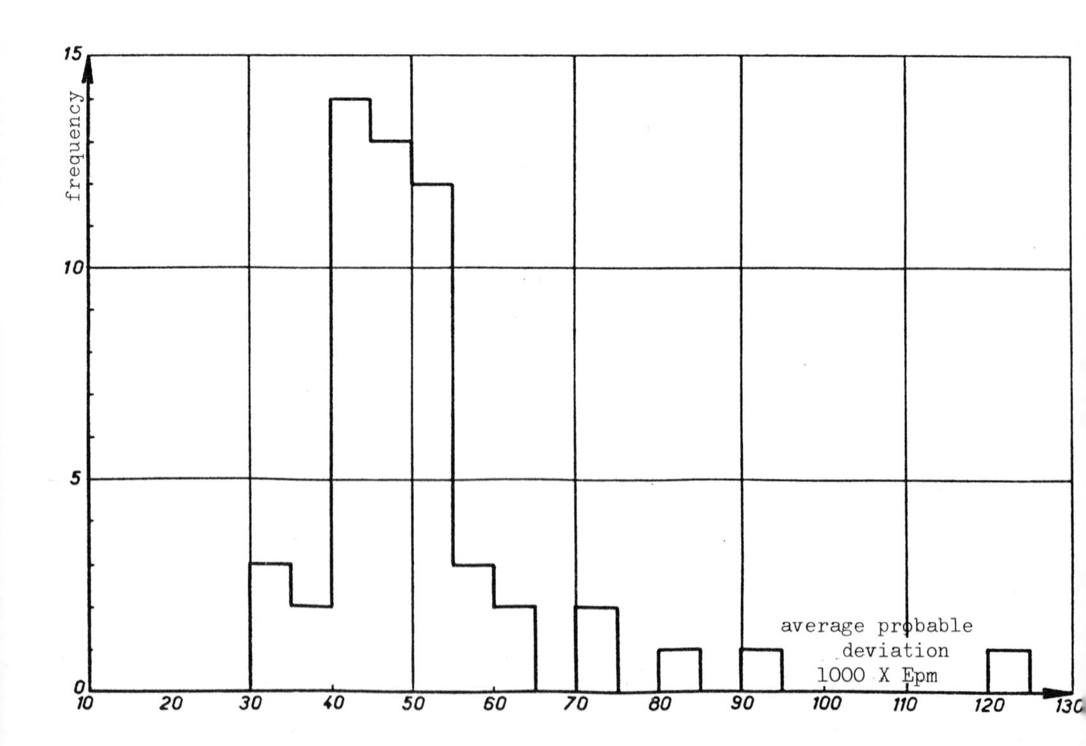

Fig. 2 - Screening Panel - Frequency of Average probable Deviation obtained during Tests.

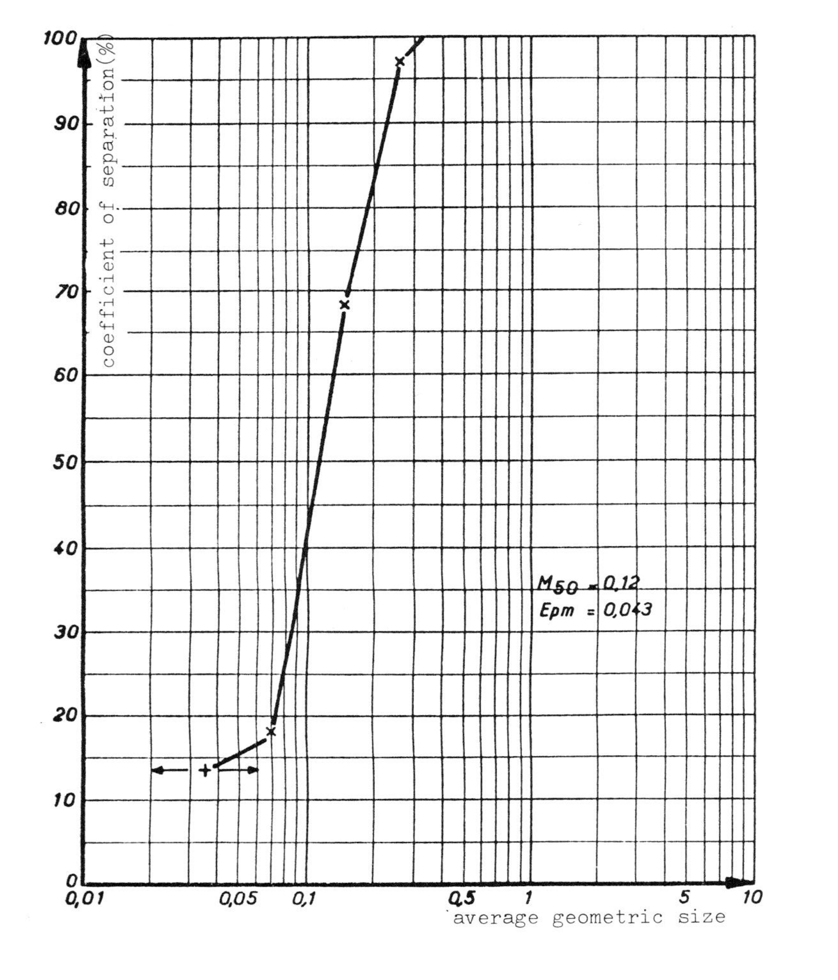

Fig. 3 - Separation Curve.

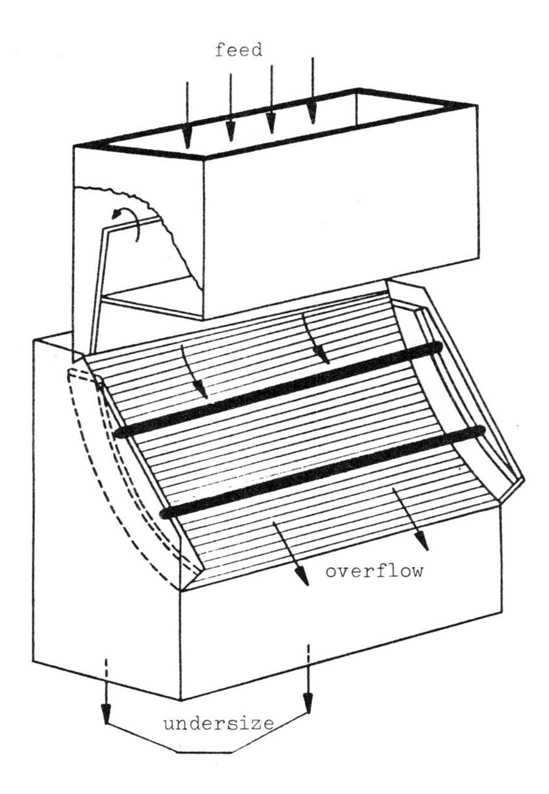

Fig. 4 - Curved Grate.

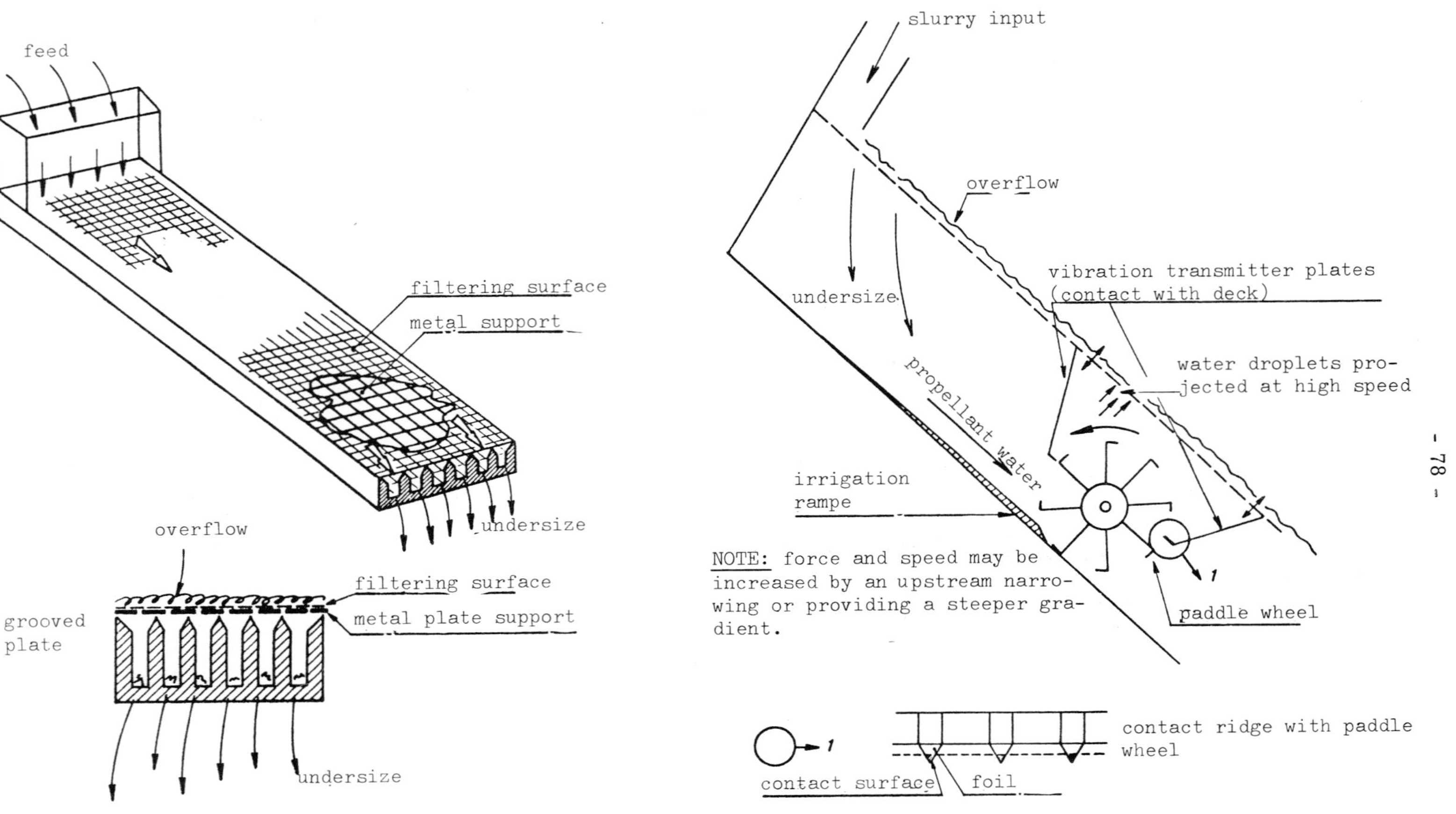

Fig. 5 - Model fitted with Device to reduce Capillarity Effects.

Fig. 6 - Layout of Paddle Wheel Vibrator

THE DEVELOPMENT AND APPLICATION OF THE ROTATING PROBABILITY SCREEN

M P ARMSTRONG

COAL PREPARATION DIVISION, MINING RESEARCH AND DEVELOPMENT ESTABLISHMENT
NATIONAL COAL BOARD

Summary

The established method of preparing power station fuel in the UK is by the part-treatment of 30 mm - 0 raw coal. This involves screening out as much of the raw fines as possible in a dry state and washing mainly the coarser size fraction thereby reducing the slurry treatment problem and minimising preparation costs. The deterioration in the condition of the run-of-mine output over many years with increased moisture, fines and dirt content resulting from widespread underground mechanisation has led to considerable difficulties in separating the fines with conventional vibrating screens because of the severe blinding of the screen decks by wet, sticky fines. The Rotating Probability Screen is an entirely new screening technique developed by the NCB Mining Research and Development Establishment which has a high resistance to blinding and can continue to remove the fines at 6 mm or below from very damp and difficult raw coals with an upper size limit of 100 mm. The new screen has the unique ability to vary the effective screening size and hence the amount of fines extracted while in operation. This new development is now available as a fully proven commercial screen which is being applied in the NCB and overseas with considerable success and economic benefit.

1. INTRODUCTION

In recent years the NCB has had annual sales of around 75 million tonnes of coal for electricity generation in the UK. This fuel is mainly supplied with an ash content between 15 and 20 percent and the majority is prepared by washing only part of the raw small coal and blending the resulting clean coal with the remaining untreated fraction to meet the product specification. There are considerable technical and economic advantages in washing only the coarse fraction of the raw smalls and keeping as much as possible of the raw fines in a dry state and thereby minimising the expensive operations of fines treatment and water clarification. However, for many years the screening of raw coal at fine sizes has been an increasingly difficult problem but the Rotating Probability Screen (RPS) developed by the NCB Mining Research and Development Establishment (MRDE) now offers a solution, with considerable economic benefits in many cases.

2. RAW COAL DETERIORATION CREATES SCREENING PROBLEM

When the part-treatment of power station smalls was widely adopted in the early 1960's the ash content of the raw smalls was such that it was only necessary in most cases to clean the fraction above 8 or 10 mm to meet the product specification and screening the raw coal at these sizes was no great problem. However the introduction of extensive coalface mechanisation during the 1960's resulted in a substantial increase in both the ash content and the fines content of the raw smalls and it became necessary to screen at 6 mm or below in order to produce sufficient washed coal, and reduce the proportion of untreated fines, to meet the required ash content of the final blend. The increased mechanisation was accompanied by the greater use of wet dust suppression methods at the coalface with a consequent increase in the moisture content of the raw coal. The increase in moisture combined with an increase in fine shale resulted in the very sticky consistency of many raw coals. Under these conditions screening at fine sizes became extremely difficult, if not impossible, due to the wet, sticky fines adhering to the screen mesh and blinding the apertures, Figure 1. In some cases fines extraction was abandoned and replaced by a 'tonnage split' in which part of the raw smalls were washed and part kept untreated with a consequent increase in the quantity of fines entering the washery system.

3. INVESTIGATION AND ASSESSMENT OF PROBLEM BY MRDE

In the late 1960's a project group was formed at MRDE to investigate the problem, to improve where possible the performance of existing screens and to develop a screening technique which overcame the problem of mesh blinding and enabled the dry extraction of fines below 3 mm from high moisture raw coal. A test screening plant was established at a production colliery and from 1970-74 many types of screens and screen decks were tested there and at other NCB collieries. These included two new screening developments, one employing flexible polyurethane screen mats which were alternatively stretched and relaxed to produce a very vigorous screening action and the other a horizontal vibrating centrifuge adapted from a dewatering to a dry screening role. Other systems tested included electrical deck heating, piano wire decks, loose-rod decks, mobile air jets and under-deck ball beating arrangements. Although a number of systems were able to produce marginal improvements in the tolerance to moisture compared with conventional screens, none was able to provide a satisfactory solution to the problem.

This work at the test plant and our wider experience in the field led to the conclusion that a radically new screening technique was needed, the basis of which should be determined according to the following criteria. It should avoid the use of fine aperture meshes and any mesh construction or deck support members which would impede the free flow of material over the screening surface. Preferably it should not be dependent on vibration either for the separating action or for material transport. The efficiency of size separation would be subordinate to the ability to tolerate a high level of moisture since continuity of operation with a limited efficiency would be of greater practical benefit for the intended application than precision of separation. The screening surface should be self cleaning so that it soon recovers its normal performance following peak levels of moisture. From commercial considerations the technique should be able to accept material up to 30 mm in size, have a capacity of at least 100 t/h and a noise level not exceeding 90 dBA. It should also be simple and robust, require minimal operator attention and be easy and inexpensive to maintain. It should be suitable for fitting either singly or in multiples into existing plants as well as new plants.

4. LABORATORY INVESTIGATION OF NEW SCREENING TECHNIQUE

In the early 1970's laboratory investigations were proceeding at MRDE into the feasibility of a new screening technique, proposed by the author, which showed promise of meeting most of the above conditions. This technique, later known as the Rotating Probability Screen (RPS) consists essentially of a horizontal circular screen deck mounted on a vertical rotating shaft equipped with a variable-speed drive. The screen deck comprises small diameter stainless steel rods radiating from a central hub and resembling a bicycle wheel without the rim. The feed material is delivered at a controlled falling velocity onto the screen deck around the central hub and is uniformly distributed around the full 360 degrees. The finer particles have a greater probability of passing through the relatively large apertures between the rods while the coarser particles tend to be retained by the deck and are discharged by the centrifugal action around the periphery. This technique has the following main features which were considered would be beneficial to the sizing of very difficult material and the avoidance of fines build-up and deck blinding.

(i) The material is screened in a free-falling dispersed condition.

(ii) The screen avoids fine aperture meshes and uses the combination of relatively large apertures and the speed of rotation to produce a relatively small effective screening size.

(iii) Since the deck comprises only radial spokes, forming outwardly expanding apertures, there are no deck elements to impede the outward flow of material and there is no opportunity for pegging by near size material which would otherwise provide a key for the build-up of damp fines.

(iv) The technique does not employ any vibrating action which could encourage fines build-up.

A further important feature of the technique is that the effective screening size and hence the proportion of underflow product is dependent on the speed of rotation of the deck and can be controlled while the screen is in operation. Increasing the deck speed reduces the effective screening size and hence the proportion of underflow. This feature is unique to this technique and makes the RPS the only instantly variable aperture screening device yet known.

The laboratory test rig was arranged for the batch testing of 100 Kg samples of 12.5 mm to 0 raw coal at feed rates up to 10 t/h. The 0.6 m diameter screen deck was made up of 2.6 mm diameter spokes and had a speed range of 90 to 266 revolutions per minute (rev/min). The laboratory tests investigated the quantitative effect of the spoke spacing, the speed of rotation and the feedrate on the size separation. The encouraging outcome of these investigations led to the decision to proceed with the design and manufacture of a pilot-scale unit with a 1.5 m diameter deck. This was installed at the MRDE test screening plant in 1974 for testing under production conditions.

5. TECHNIQUE PROGRESSES TO PROTOTYPE SCREEN

Development of the screen continued on the pilot-scale machine and a special feed distribution system was incorporated to ensure a uniform circular distribution of the feed material to the screen deck. The machine was able to handle feedrates of up to 50 t/h with acceptable levels of efficiency. The screen deck remained free from 'blinding' with high moisture coals and although some fines adhered to the spokes at peak moistures, this was not sufficient to bridge the apertures and the deck gradually cleared when the moisture decreased.

The success of the pilot-scale unit led the NCB to proceed early in 1976 to a pre-production prototype unit capable of treating 100 t/h. The detail design and manufacture of two prototype machines to an NCB specification was undertaken under contract by a private company. The units had a 2.4 m diameter deck originally comprising 360 stainless steel spokes of 5 mm diameter and later reduced to 240 spokes of 6 mm diameter. The deck speed was variable from 25 to 100 rev/min. They were equipped with a mechanical feed distribution system and the products were collected in steep-sided chutes; the inner underflow chute having a single outlet and the outer overflow chute dividing into two outlets to limit the headroom requirements. These machines were installed during 1977/78 at Bevercotes and Lea Hall Collieries for the extraction of fines below 6 mm from 19 mm - 0 and 25 mm - 0 raw coal respectively in the preparation of part-treated power station fuel.

6. DESIGN OF PRODUCTION SCREEN

6.1 Overall Design

In early 1978, six months after the successful commissioning of the prototype screen at Bevercotes, it was felt that sufficient operating experience had been gained to proceed with the design and manufacture of the final production screen. This was again undertaken by a private company to NCB instructions. Three machines were to be manufactured and installed during 1979 as industry demonstration units with the purpose of demonstrating the practicability, reliability and effectiveness of the RPS in extracting fines from high moisture raw coal and promoting the wider application of this development within the NCB.

The production screen, Figure 2, is a free-standing, composite unit incorporating its own feed presentation and product collecting systems. It has an overall height of 3.7 m, occupies a floor area 3.4 m square and weighs 9 tonnes. It has three drive units with a total installed power initially of 17 kW and later increased to 19 kW. The design, illustrated in Figure 3, was based on the prototype screen, having a 2.4 m diameter deck, but took into account the manufacturing, installation, operating and maintenance experience gained with the prototype unit.

6.2 Feed Presentation and Screen Deck

The feed is delivered down a vertical feed chute onto the rotating feed table and is initially contained between the overlapping sections of the stationary spiral plough. The table rotates at 12 rev/min and the feed is displaced uniformly from the edge and falls 300 mm onto the rotating screen deck. The screen can be fed at an independently controlled rate or in the case of a multiple installation it can be choke-fed from a distribution scraper or shallow hopper in which case an adjustable profile plate would be fitted at the outlet of the feed chute to control the bed-depth on the rotating table and hence the feed rate.

The screen deck is composed of 12 standard interchangeable segments each comprising 20 stainless steel spokes of 6 mm diameter. The spokes are a drive fit into undersize holes in a polyurethane block moulded to a steel segment plate and they can be individually replaced. The entry of the holes in the polyurethane is tapered to produce a gradually tightening grip on the spokes and thereby avoid a high stress concentration at the point of entry.

The deck segments attach to the hub plate which is secured to the screen shaft. The shaft is rigidly coupled to the vertical output shaft of an 'agitator-type' worm reduction gearbox and there is no lower bearing. The drive is from a 7.5 kW motor through an eddy-current variable speed coupling with control unit arranged for both local and remote manual control and speed indication and it can also accept a 4-20 mA signal for automatic control. The deck speed is infinitely variable up to 120 rev/min although the normal operating range is 40-80 rev/min.

6.3 Product Collection and Discharge System

A major departure from the prototype design was in the product collection. Both the finer product passing through the deck and the coarser product discharged around the periphery are collected on an annular rotating table with a cylindrical division, corresponding to the edge of the deck, to keep them separate. The discharge table rotates at between 4 and 6 rev/min and the fines are ploughed off the inside of the table into a single inner discharge chute and the coarser material is ploughed off the outside into a single outer chute. With this product collection and discharge system almost all the surfaces which are in contact with the products are scraped each revolution of the table thereby avoiding any build-up of wet fines. The exception to this is the inside of the cylindrical screen casing. To avoid build-up here a soft rubber curtain is suspended around the inside of the casing and any slight build-up of fines soon breaks away because of the flexibility. The curtain also prevents the breakage of the coarser material thrown from the screen deck as well as reducing noise.

Incorporating the rotating discharge table into the production design reduced the overall height by almost 1.5 m, and together with single discharge points for each product, it could be of considerable advantage in installing the screen in existing plants. Should there be a need for a larger capacity screen then it would permit the diameter to be increased without any appreciable increase in the headroom requirements. Advantage was taken of the circular configuration of the RPS to provide the maximum flexibility of layout particularly with respect to existing plants. Provision was made in the design for 16 possible variations in the relative positions of feed and product chutes to suit existing conveyor systems and this was achieved with only two basic casing designs.

6.4 Environmental and Maintenance Features

Regard for the working environment in coal preparation plants and the advantages of a low maintenance requirement were also taken into consideration in the design of the production unit. The total enclosure of the screen helped to reduce further the low noise level of the rotating system in addition to preventing spillage and dust. All mechanical components were generously designed or selected from proven proprietary equipment for long maintenance-free running. Provision was made for access to the few parts which require regular inspection and occasional attention. The ease of replacement of the screen deck was given particular attention. The speed of rotating components was kept as low as possible to minimise stress and wear. All surfaces subject to wear, from the material being handled, were where practical provided with replaceable linings.

7. INDUSTRY DEMONSTRATION INSTALLATIONS

The first two industry demonstration installations of the production RPS were made at Markham Colliery, NCB North Derbyshire Area and Hickleton Colliery, NCB Doncaster Area. They were commissioned during July and August 1979. Both units have been in continuous use and have demonstrated the high resistance to blinding, reliability of operation and low maintenance requirements of these new screens. At Markham six 4.9 m x 1.85 m vibrating screens were being employed to extract fines below 6 mm from 400/450 t/h of 30 mm - 0 raw coal. These screens had been fitted with 'loose-rod' decks to overcome the problems of deck blinding caused by high moisture and clay fines in the raw coal but they were having very limited success and required considerable maintenance. The RPS proved capable of taking over the duty of two of the existing vibrating screens. Prior to the installation of the RPS at Hickleton some fines were being extracted through 6 mm stainless steel wedge wire panels on the bottom deck of the double-deck, raw coal jigging screen. The high level of moisture in the raw coal caused frequent blinding of this deck and regular manual cleaning was necessary. As a result of these difficulties less fines were extracted and more fines entered the washing system causing additional slurry recovery problems.

At this stage the top size of the raw coal fed to the RPS had been limited to 30 mm because of the screen deck construction. With many NCB plants treating 100 mm or 125 mm - 0 raw coal on a single Baum jig this would require a secondary screening operation at around 25 mm prior to the RPS. The third industry demonstration unit was therefore used to explore the possibility of handling a 100 mm - 0 raw coal feed by incorporating a scalping deck above the sizing deck as shown in Figure 4. The scalping deck comprising 120 stainless steel rods of 12 mm diameter was supported from the underside of the circular feed table and rotated with the table at 12 rev/min. To compensate for the lower speed the scalping deck was angled downwards at 10 degrees to assist the outward travel of the oversize material.

This third production screen was installed at the end of 1980 at Darfield Colliery, NCB Barnsley Area in place of one of three vibrating screens fitted with loose-rod decks for extracting fines from 100 mm - 0 raw coal. The scalping deck limited the top size of the feed to the sizing deck to about 30 mm and the overflow from both decks was collected as a single product. A further effect of the scalping deck was to increase the radial dispersion of the feed material onto the sizing deck and although this reduced the concentration of feed near the hub it necessitated the strengthening of the sizing deck. This was successfully achieved by extending the 6 mm spokes and bending the ends downwards to locate in undersize holes in a polyurethane strip. These strips on adjacent deck segments are joined to form a continuous band which extends around the outside and below the rim of the product dividing cylinder and does not therefore impede the discharge of the overflow product from the deck or cause any fines build-up. This design of deck can be used with advantage on a single-deck screen where the size of the feed is 25 mm or above.

A further innovation of the Darfield installation was that the screen was choke-fed from a distributing scraper conveyor with an adjustable gate at the discharge of the feed chute to regulate the bed depth on the feed table and hence the feed rate. This feed arrangement produced a very even bed of material on the feed table and consequently a good circular distribution to the screen deck. Despite the very damp and difficult nature of the raw coal feed there were no problems of build-up or blockages in the feed chute with this system.

8. OPERATING EXPERIENCE WITH PRODUCTION RPS

All three installations of the production RPS demonstrated quite convincingly that the problems we set out to overcome with this new screen have been successfully solved. It has been clearly shown that the sizing technique employed by the RPS, in which the apertures are several times larger than the effective screening size, is capable of handling high moisture coals without blinding and will still continue to extract fines from very difficult materials. The feed and product handling arrangements have been shown to be equally capable of handling very damp, sticky material, thereby enabling the capabilities of the screen to be fully utilised. The final production screen has proved to be a practical machine capable of continuous and reliable production operation and with a low maintenance requirement. The screen deck and the rubber curtain on the inside of the casing will handle approximately 0.5 million tonnes of feed before needing replacement. Replacement of the screen deck is both simple and quick; a complete deck being changed in less than two hours by plant maintenance staff. The ploughs for the feed and products which have been faced with ceramic tiles will handle over 1 million tonnes of feed.

The low mechanical maintenance requirement is attributable in part to the absence of any vibrating action which together with the total enclosure of the screen is also responsible for the low noise level. The fines screening section at Darfield, being housed in a separate building from the main plant, allowed comparison of screen noise levels without the background noise of the plant. The RPS screen gave readings of sound pressure levels of 72 to 78 dB(A) compared with 99 dB(A) for the vibrating screens fitted with loose-rod decks.

The prime objective in the development of the RPS was a non-blinding screen with screening efficiency as a secondary, but important, consideration. The level of screening efficiency so far attained, must be viewed with due regard to the other advantageous features of this technique. The level of underflow efficiency ranges from 75% at an effective separating size of around 6 mm to 60% at 3 mm and is regarded as sufficient to provide operational and financial benefits at many collieries preparing blended power station fuels by the part-treatment of raw coal. Although designed for a capacity of 100 t/h of 30 mm - 0 raw coal, experience has shown that the production screen can handle 150 t/h, providing neither product exceeds 100 t/h, without any significant drop in efficiency. Compared with conventional screens the capacity per unit area of screen deck is very high with a feed capacity of 60 t/h per m^2 and an underflow extraction capacity of 40 t/h per m^2. The incorporation of the scalping deck allows the feed size to be increased to 100 mm and the capacity to 200 t/h.

9. AUTOMATIC CONTROL OF SCREENING SIZE

The unique ability of the RPS to vary the effective screening size instantly while operating on load has been demonstrated to be of considerable benefit in preparing part-treated power station fuel. With a 240 spoke deck the effective screening size can be varied from 6 mm at 50 rev/min to 2 mm at 70 rev/min and with a typical 25 mm - 0 raw coal feed this would result in the fines extraction varying from 50% to 20% of the feed. This facility is capable of further exploitation by enabling the automatic control of fines extraction according to the ash content of the raw coal in the preparation of power station fuel as shown in Figure 5.

The fines extracted from the raw coal by the RPS are fed direct to one bunker and the coarser fraction from the screen is washed either by Baum jig or dense medium and the resulting clean coal is fed to a second bunker. The levels in both bunkers are continuously monitored and the signals taken to the screen speed control unit. The power station fuel is produced to the required specification by blending the appropriate proportions of washed and untreated components, the latter being added by an NCB-developed blender-feeder which regulates the feedrate according to the bulk density and hence the ash content. A final check on the product quality can be made with an on-line ash monitor.

When the ash content of the untreated fines increases less fines are added to the blend and the bunker level rises. The screen speed will automatically increase to reduce the screening size, extract less fines and produce more overflow for washing. Conversely when the ash content of the untreated fines decreases more fines are added to the blend, the bunker level falls and the screen speed reduces to extract more fines. In this way the system responds to variations in the ash content of the raw coal and the screen speed is automatically controlled to make the blend constituents available in the required proportions. The system would also ensure that only the minimum amount of raw coal was washed to meet the blend requirements and therefore the quantity of fines entering the washing system was kept to a minimum.

10. COMMERCIAL EXPLOITATION

Following the successful commissioning of the first production screens in 1979 a licence was granted by the NCB to Magco Ltd, a company in the Babcock International Group, to manufacture and market the Rotating Probability Screen throughout the world and they adopted the registered trade name of 'Ro-Pro'. More recently sub-licences have been granted by Magco to the Gundlach Division of Rexnord in North America and to Vickers

Australia Ltd.

Since 1980 a further thirteen production screens have been installed on existing coal preparation plants at eight NCB collieries. The applications have all been in concerned with the part-treatment of power station fuel and include three double unit installations and one triple unit installation of double-deck machines. These installations have mainly replaced existing vibrating screens which were unable to handle difficult, high moisture coal. They have confirmed the resistance to blinding, reliable operation and low maintenance requirements of the Ro-Pro screen. The facility to control the extraction of fines has been used to good effect on most installations to improve the consistency of the power station blend. Provision for automatic control of the screen speed according to the level in the untreated fines blending bunker has been incorporated on the latest installation and trials with the control system are continuing. Most of the installations have produced substantial financial benefits for the NCB both by increased proceeds and reduced costs of slurry treatment. One single unit installation costing £64,000 benefitted the colliery to the extent of £470,000 per year.

The Ro-Pro is finding application in overcoming difficult screening problems overseas and units are operating on coal in the USA, Australia, Sardinia and China. The application to other materials is so far limited to a single unit installation in Eire for separating fibrous roots from peat prior to briquetting. However, interest in this screening technique has been shown by a wide range of industries including alumina/bauxite, gypsum, rubber reclamation, brewing, flour milling, sugar beet, sewage, and the screening of domestic refuse.

11. ACKNOWLEDGEMENTS

The author wishes to thank Mr T L Carr, Head of Mining Research and Development, NCB, for permission to publish this paper. The views expressed are those of the author and not necessarily those of the National Coal Board. The development of the Rotating Probability Screen from January 1975 to June 1978 was undertaken with financial aid from the ECSC.

12. BIBLIOGRAPHY

ARMSTRONG, M P; ATKINSON, A; JENKINSON, D E; ROBERTS, J and TURNER, A. "The Dry Extraction of Small Coal and Fines from Moist Raw Coal" 6th International Coal Preparation Congress, Paris, March 1973.

ARMSTRONG, M P. "Dry Fines Screening" Colliery Guardian, Vol 223, No 5, May 1975, pp 162-170.

"Extraction of Fine Particles from Raw Coal" Final Report on ECSC Research Project 6220-EA/8/801, MRDE, 1979, 78 pp.

"Rotating Probability Screen" Editorial Article based on MRDE Report presented at Hickleton Colliery RPS Demonstration, World Coal, August 1980, pp 29-31.

SHAW, S R. "The Controlled Extraction of Dry Fines from Raw Coal" Gluckauf-Forschungshefte 43 (1982) October, pp 218-221.

JENKINSON, D E. "Rotating Probability Screens" Mining Congress Journal, July 1982.

SHAW, S R. "The Rotating Probability Screen - a new concept in screening", Mine & Quarry, October 1983, pp 29-32.

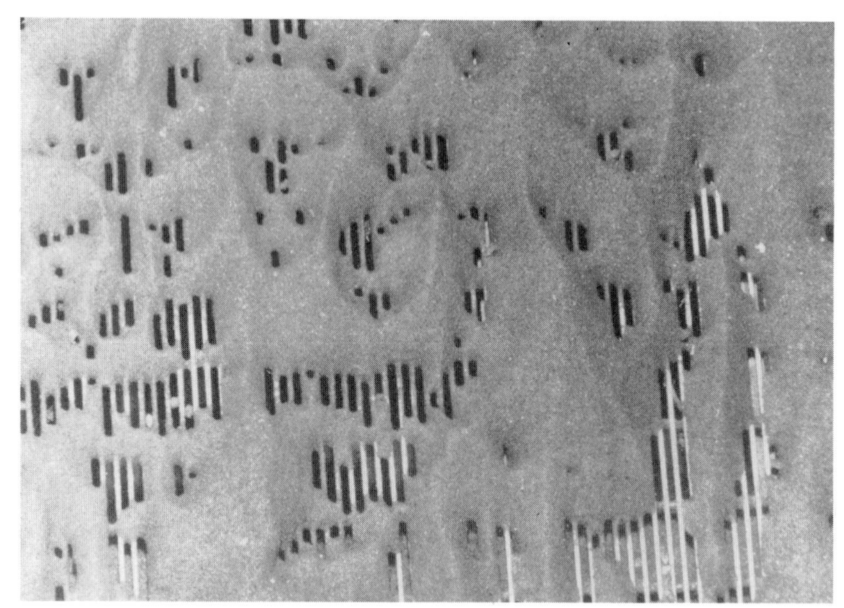

Figure 1. Wet, sticky coal causes rapid 'blinding' of conventional vibrating screen

Figure 2. Production RPS allows good maintenance access to the deck

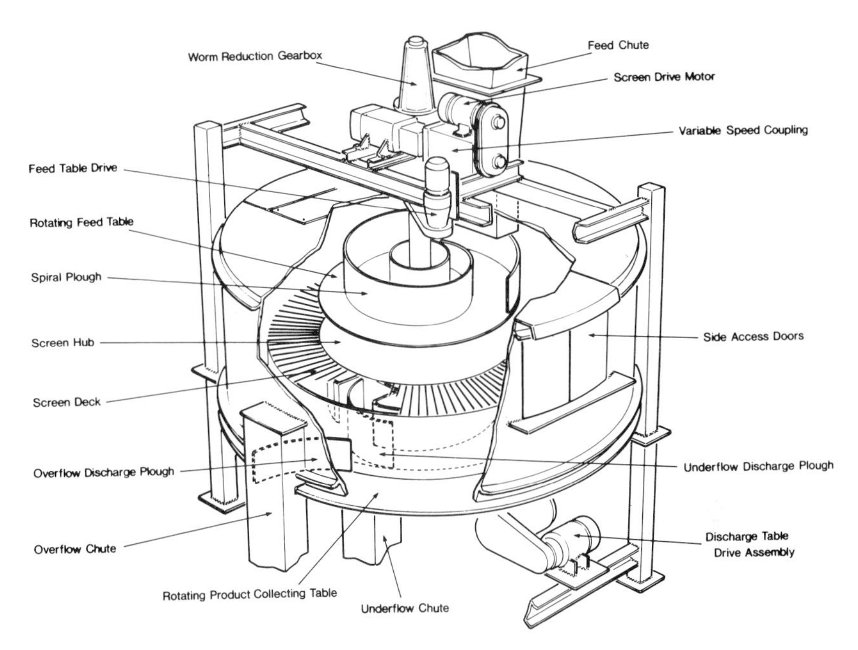

Figure 3. Production RPS incorporates feed distribution and product collection systems

Figure 4. Production RPS with scalping deck handles 100 mm - 0 feed at Darfield Colliery

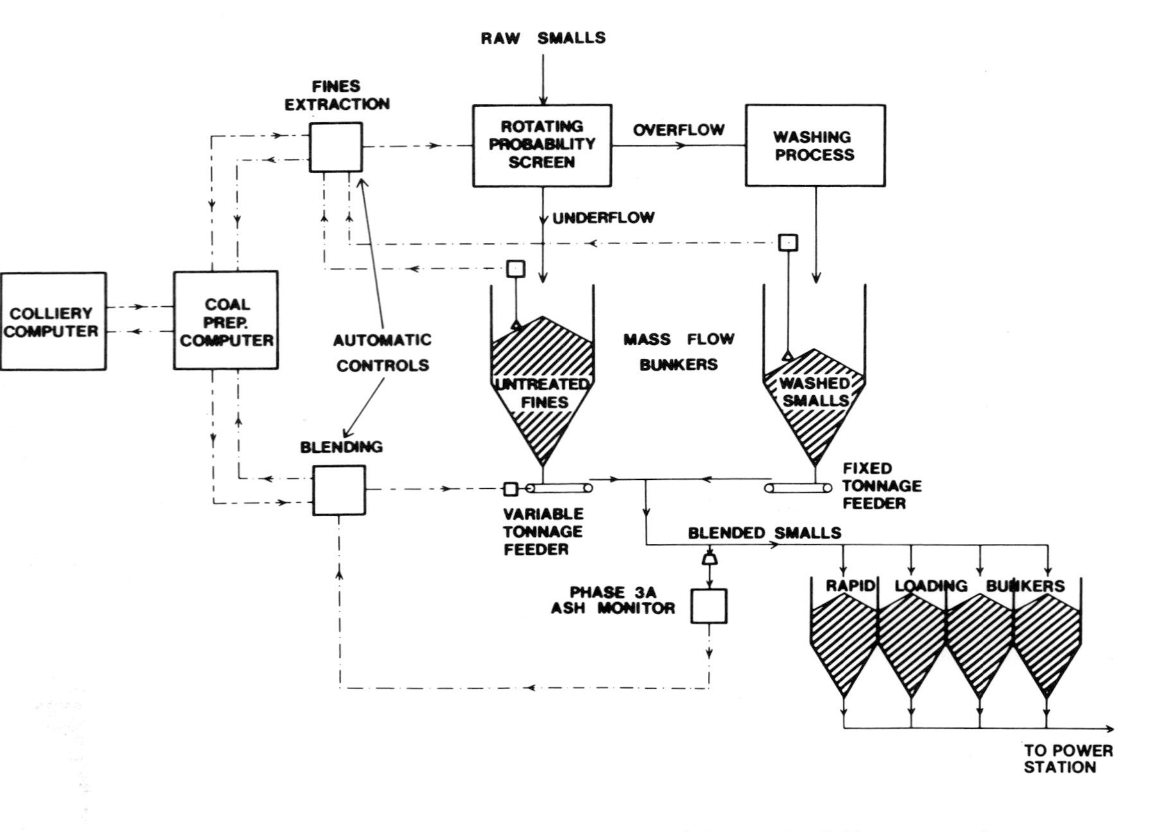

Figure 5. Proposed automatic speed control of RPS will enable fully automatic preparation of power station fuel

RECENT DEVELOPMENTS IN THE DEWATERING OF FINE AND SUPER-FINE COAL PRODUCTS

W.P. ERDMANN
Scientific Member
Bergbau-Forschung GmbH
Coal Preparation Division

Summary

The observance of a water content conforming to market trends in the fine and ultra-fine coal products is becoming increasingly problematical, because the greater market demand for low-sulphur high-grade coal for economic and environmental reasons is causing separation to be extended more and more towards the ultra-fine range. To avoid the expensive and environmentally important thermal drying, efforts are being made in the coal preparation sector to counteract this trend both by improvement and further development of already known dewatering methods and equipment and testing those not previously used in coal preparation. Some partial successes have already been achieved in past years, even though - in somewhat general terms - the solids yield of the centrifuges and residual water content of the filters are apparently still in need of improvement. Further improvement potential is anticipated in division of the particle range below about 10 mm into more than the hitherto customary two partial ranges and the separate dewatering of these narrow ranges in specially designed equipment. An important contribution was made with the introduction of the solid bowl centrifuge into the dewatering of ultra-fines, even though after-treatment of the ultra-fine concentrate solids has not yet been finally solved. According to the investigation results so far available the application of pressure filtration to the dewatering of ultra-fine coal slurries is expected to be particularly successful. The aim of the present and future investigations is revision of the overall concept of dewatering in the particle size range below about 10 mm with inclusion of the ultra-fine residual solids discharged in the concentrate or filtrate of dewatering equipment.

1. INTRODUCTION

In the German coal mining industry fine coal is usually dewatered in only two process stages, viz. dewatering of fines from about 10 to 0.5 mm in jigging screen centrifuges and of ultra-fines below about 0.5 mm on rotary suction cell filters. In view of the anticipated full sizing of the ultra-fines in the medium and long term both for reasons of market demand for high-grade coal and the pressure to reduce the sulphur content this method and the dewatering equipment used hitherto will give rise to an appreciable increase in the water content of the fine coal, which in some cases already does not meet market requirements. Application of the expensive and environmentally important thermal drying - at least for partial quantities and partial products - will thus be unavoidable, if a significant improvement in the mechanical dewatering of fines and ultra-fines cannot be achieved by suitable measures.

A series of investigations, some of which produced highly favourable results and have meanwhile led to the use of various new or developed machines has already been conducted in this connection in past years. It can certainly also be counted as a success of the previous efforts that despite the deteriorations in the unwashed coal properties and the already introduced extension of the sizing in the ultra-fine range it has been possible to comply with the fine coal water content required by the market (1 to 4).

Extension of the sizing in the ultra-fine range is, of course, accompanied by an increase in the yield of ultra-fine tailings, which are particularly difficult to dewater because of their clay content. The widespread dumping of these tailings in ponds in the past is significantly less appropriate nowadays and in future will be possible only in exceptional cases because of the lack of suitable sites and also for safety and environmental reasons. The further development and improvement of existing and the testing of new dewatering equipment are thus particularly important for dewatering the ultra-fine tailings at the lowest possible cost to a residual moisture content required for mixed dumping with coarser tailings or use elsewhere.

The investigations into improvement of the dewatering of fines and ultra-fines must be continued in order to create an adequate basis for revision of the concept of overall dewatering, which also takes into account in particular the ultra-fine residual solids discharged in the filtrate or concentrate of dewatering equipment and thus particularly difficult to dewater.

2. DEWATERING OF FINE COAL

Nowadays bucket conveyors, jigging screens or grizzlies are used for preliminary dewatering of washed fine coal. Preliminary dewatering in bucket conveyors is practised only in a few cases because the important problem apart from preliminary removal of water, viz. keeping the ultra-fines out of the centrifuge feed as far as possible and uniform feeding of the centrifuges, is solved only inadequately. However, jigging screens and in particular grizzlies, i.e. fixed inclined screens can perform this task satisfactorily only if they are sufficiently large and uniform distribution of the feed material is ensured.

Apart from special individual cases further investigations into preliminary dewatering are at present unnecessary.

The main dewatering of the coal now takes place almost exclusively in jigging screen centrifuges, whereas the previously widely adopted screw screen centrifuge is used only in a few special cases. So far only a few tests have been conducted with the timbler centrifuge. The following comments can be made on the current state of development of individual machines:

2.1 Jigging screen centrifuge

The jigging screen centrifuge operating with centrifuging factors between 80 and 100 g is now the standard machine for dewatering washed fine coal fractions between about 10 and 0.5

mm. The prerequisites for high throughput rates and easy access for repair and maintenance were created by design improvements particularly by the change from vertical to horizontal construction. Jigging screen centrifuges are currently built with basket diameters up to 1500 mm and rated outputs up to 400 t/h (Figure 1). However, endurance tests on a machine of this type have shown that with this basket diameter small quantities of adhering material already lead to considerable imbalance and the resulting trouble and downtimes. The machines with a basket diameter of 1300 mm and a rated output of 250 t/h now in widespread use thus apparently constitute an optimum size. However, it should be taken into account that the rated outputs of the machines specified by the manufacturers really only indicate the absorption capacity and satisfactory process engineering results are achieved only if the feed rate is limited to about 65% of the so-called rated output.

Whereas yield of solids with values around 98% is satisfactory, the problem of "spillage water" in vibrating screen centrifuges has not yet been solved, because part of the concentrate is wrongly discharged with the solid. To eliminate this defect some centrifuges were provided with a so-called labyrinth seal. Tests and investigations on these machines are still in progress.

A final evaluation of the stepped basket (Figure 2) designed to rearrange the solids, which have already largely undergone preliminary dewatering in the feed section of the centrifuge, and thus achieve better dewatering is likewise not yet possible.

2.2 Screw screen centrifuge

Screw screen centrifuges, which operate with centrifuging factors between 120 and 180 g and largely dominated the field of mechanical dewatering in the Fifties, were replaced almost completely by jigging screen centrifuges because of their limited throughput, susceptibility to trouble and difficult access for repair and maintenance work due to the vertical construction, although they permit up to 1% lower water content, but with up to 3% smaller solids yield. Meanwhile a new horizontal screw screen centrifuge which could possibly also be of interest to coal preparation, is in use in the chemical industry and allegedly offers the following advantages compared to the previous type (Figure 3):

- horizontal construction, hence separation of drive and dewatering section and easier maintenance and repair,
- higher throughput rate due to change-over to 1200 mm basket diameter,
- use of new improved wedge-wire screen baskets, hence less susceptibility to trouble and longer intervals between maintenance,
- dewatering of thick layers by redesign of the feeder screw, hence higher discharge of solids.

The first tests on a centrifuge of this type in coal preparation will be conducted this year.

2.3 Tumbler centrifuge

A further machine type operating with a higher centrifuging factor between about 180 and 300 g is the timbler or gyrating centrifuge (Figure 4). On this machine the conveying momentum is achieved by superimposition on the bowl rotation of a forced precession or timbling movement of the drum axis inclined by a few degrees. The machine characteristic can be varied relatively easily by changing the centrifuging factor and tumbling angle, so that the machine can be used for different tasks. Tests have recently been conducted in two preparation plants with this machine; the results so far obtained can be summarised as follows (5):

The tumbler centrifuge produces a final water content up to 2% lower than the jigging screen centrifuge in the dewatering of washed fine coal below 6 mm. The throughput rate is 35% higher than that of a jigging screen centrifuge with the same basket diameter.

By contrast the solids yield with values around or below

90% is still unsatisfactory. The so-called "Sprayed particles" i.e. misdirection of already dewatered solids into the concentrate, is one of the factors responsible for this situation. Meanwhile some design modifications, which are currently still being tested, have been undertaken.

The originally heavy wear was reduced by reduction of the centrifuging factor to about 190 g, armouring and fitting of guide ribs in the generatrix direction. At present a new basket, in which the actual screen basket is separated from an "impact container" in the feed zone, which fills with coal and acts as a feed cushion, is being tested. So far more than 100 000 t have passed through this basket without occurrence of excessive wear.

Because of its substantially higher capital and operating costs compared to the jigging screen centrifuge the tumbling centrifuge probably hardly comes into consideration for the relatively easily dewatered fine range between 10 (6) and 0.5 mm, especially since no important process engineering results can be achieved in this particle size range - at least according to the present state of knowledge. However, it is probably interesting for the coarse slurry range, i.e. for particles smaller than about 3.0 mm, now that the still existing problems can apparently be solved.

3. DEWATERING OF COARSE SLURRY

Higher centrifuging factors and longer residence times than for the fine coal are required for dewatering coarse slurry in the particle size range below about 3.0 mm. Because of their machine characteristic timbler, pusher, screw screen and solid bowl screen centrifuges come into consideration in this case. The current situation is as follows:

3.1 Tumbler centrifuge

As already described above, the problems with regard to wear were largely solved in the tests on the tumbler centrifuges in the dewatering of fines. If the efforts to improve the yield of solids are successful, this machine will probably also come into the question for dewatering of coarse slurry, which would have to be checked by a test under operating conditions.

3.2 Pusher centrifuge

Pusher centrifuges operating with centrifuging factors between 400 and 500 g were used in a number of treatment plants for different applications in the past. Quite good process engineering results were achieved in some cases. However, the sensitivity of the machine to fluctuations in the feed quantity and quality proved to be disadvantageous. The change-over to a two-stage machine type also did not produce the anticipated success, so that variations in the preliminary concentration of the feed still lead to imbalance, operating trouble and downtimes. According to the available knowledge constant preliminary concentration of the feed preferably to more than 700 g/l is required to ensure acceptable operating conditions.

However, the pusher centrifuge was highly susceptible to trouble even with high concentration - tests by Saarbergwerke on the subsequent dewatering of filter sludge (6,7). Hence further engineering developments and improvements by the machine manufacturers are required before further use under practical conditions.

3.3 Screw screen centrifuge

If the improvements for the horizontal screw screen centrifuge specified by the manufacturer and already listed above are in fact achieved, this machine will probably also be of interest for the coarse slurry range at least in the cases where the proportion of very fine and ultra-fine particles is not too high.

3.4 Solid bowl screen centrifuge

Solid bowl screen centrifuges, which are suitable primarily for the ultra-fine range because of their high centrifuging factors of 500 g and more, may also come into consideration for the coarse slurry range under certain circumstances, though probably only in cases where it can be anticipated on the basis of the particle size distribution that the solids quantity contained in the concentrate is not too large, i.e. the ultra-fine proportion in the centrifuge feed is not too high.

4. DEWATERING OF ULTRA-FINES

Rotary suction cell filters dominated concentrate dewatering and - insofar as dumping in ponds was not possible - chamber filter presses dewatering of tailings for many years in the ultra-fine range below about 0.5 mm. In recent years several new machines have been introduced into coal preparation, at least some of which are firmly establishing themselves.

4.1 Centrifuges

4.1.1 Solid bowl screen centrifuge

In practice a combination of solid bowl and screw screen centrifuge (Figure 5), the solid bowl screen centrifuge has produced good results in various tests and use under practical conditions in the dewatering of flotation concentrate (8,9). The originally heavy wear was significantly reduced by improvements such as lining the machine parts exposed to particularly high stresses with ceramic material, so that sufficiently long basket and screw lives are achieved.

Whereas the engineering aspects of operation of the solid bowl screen centrifuge are unproblematical, and are favourably evaluated, the further processing of the concentrate may cause considerable difficulties. If the solids quantity contained in the concentrate is not too large because of the particle size distribution of the centrifuge feed and combined operation with vacuum filters is carried out, the concentrate can be fed to the latter. Vacuum filters process the ultra-fine concentrate, although an increased water content in the filter cake must be tolerated.

Further treatment of the concentrate is substantially

more problematical with higher ultra-fine proportions in the centrifuge feed and a correspondingly higher solids content in the centrifuge water, particularly if centrifuges exclusively are used for the dewatering.

Figure 6 shows the effect of the ultra-fine proportion in the centrifuge feed on the dewatering result. It is clear that with ultra-fine fractions below 0.063 mm over 45 - 50% by weight the water content increases sharply and the solids discharge falls appreciably, i.e. the overall dewatering result greatly deteriorates. For the sake of completeness it should be mentioned that these tests were conducted with a machine of type SVS 1100 x 3300. The feed rate was 20 t/h (wf), the pond height was 85 mm and the centrifuging factor 503 g.

With dewatering in solid bowl screen centrifuges a sorting effect occurs, so that the ultra-fine solids of the concentrate may have an ash content between about 25 and 40% by weight depending on the raw material conditions. This material often cannot be incorporated in high grade end products for quality reasons, while on the other hand if it were to be discarded into the preparation tailings unjustifiably high coal derivative losses would result. For these reasons further investigations into dewatering and - at least in some cases - also into subsequent sizing of the concentrate of solid bowl screen centrifuges as well as dewatering of the resulting ultra-fine concentrates and tailings are currently being conducted.

4.1.2 Solid bowl centrifuge

Solid bowl centrifuges are used for dewatering flotation tailings in some treatment plants (10). Because of the relatively high final water content of up to 40% by weight and the pasty consistency of the centrifuge cake flotation tailings dewatered by solid bowl centrifuges lead to fouling and blockage of belts, chutes and bunker outlets, so that solid bowl centrifuges come into consideration only if the dewatered flotation tailings are admixed with the fine and coarse washed tailings, and this probably also only in cases where the proportion of flotation tailings in the total washed tailings is extremely low.

An additional disadvantage of the solid bowl centrifuge is the substantial flocculant requirement, the residual concentration of which in the circulating water may have an adverse effect on the selectivity of the flotation.

4.1.3 Combination of solid bowl and pusher centrifuges

On the assumption that the operating conditions in the solid bowl screen centrifuge represent a compromise for the solid bowl and screen section, it was thought that concentrate dewatering could be improved by combination of solid bowl and pusher centrifuges, because both units can then operate under optimum conditions.

However, a test on this machine combination lasting several months was not very successful not least because of the already mentioned sensitivity of the pusher centrifuge to fluctuations in the feed quantity and quality.

4.1.4 Special centrifuges

Despite good dewatering results special centrifuges such as the flat-bottom centrifuge have so far not been adopted in coal preparation, because they do not meet requirements from the engineering and control point of view.

4.2 Filters

4.2.1 Vacuum filter

Drum and disc filters are now part of the standard equipment in the dewatering of ultra-fines. In the past preferably disc filters were installed, particularly when extending existing washery capacities, because they provide larger filter areas than drum filters with the same area and volume requirement. However, drum filters will probably by more widely used in future with the increasing fineness of the feed material, because they offer greater scope for improvement of cake removal and for cleaning the filter fabric, e.g. by removal of the cloth (11).

The initial use of a vacuum belt filter for concentrate dewatering has been described on various occasions (12,13). Hence to summarise it will only be pointed out that the expectations with regard to the process engineering result - which admittedly were extremely high - were not fully met. In addition to the slightly lower residual water content compared to drum filters, however, the vacuum belt filter has the advantage that fluctuations in the quantity and quality of the feed material can be handled within a wide range even though they have some effects on the water content of the filter cake. A disadvantage of this filter is undoubtedly the area requirement dictated by its design, which makes subsequent installation in an existing washery building virtually impossible.

The vaporising method has so far been used on an industrial scale with drum and belt filters (13 to 15). Despite good process engineering results it has not yet been able to establish itself at least for drum filters. The reasons are firstly the need to use special materials and, secondly, greater maintenance.

Furthermore, a high proportion of ultra-fines and the resulting low porosity of the filter cake may make it difficult to apply the method. In addition the reduction of the final water content achieved by vaporising is limited.

4.2.2 Pressure filters

With chamber filter presses a well-proven system is available for the dewatering of ultra-fine slurries, e.g. flotation tailings. A further development in the form of a membrane chamber filter press, with which highly satisfactory results have allegedly been achieved in the dewatering of ultra-fine solids, has been undertaken in Great Britain (16). This press, in which an additional pressure increase is achieved by pneumatically operated pressure pads, could be of interest for the dewatering of ultra-fine solids such as the products of pnummatic flotation or also of the concentrate of solid bowl screen centrifuges, if conventional chamber filter presses were to be inadequate for this purpose.

Meanwhile filter belt presses have likewise been adopted in the dewatering of flotation tailings. To improve profitability, possibilities of increasing the machine availability and in particular reducing the flocculant requirement still have to be investigated (17 to 19).

The introduction of filter belt presses into the dewatering of flotation tailings has at least caused the manufacturers of chamber filter presses to make improvements in the design and reconsider their prices.

As the pressure differential of about 0.8 bar representable under practical conditions by vacuum filtration no longer permits the surmounting of the capillary pressure required for satisfactory dewatering particularly with an increasing ultra-fine proportion, application of the pressure filtration not only to ultra-fine tailings containing clay, but also to ultra-fine coal slurries is currently being considered.

In a trial lasting about one year the prototype of a continuous pressure filter was investigated in the concentrate dewatering of a preparation plant in the Ruhr coal field (20,21). An extremely good process engineering result was achieved with residual water contents of 16 - 17% by weight and a specific filter output of 500-600 $kg/m^2.h$, because the corresponding values for the vacuum filters were only 21 - 22% by weight water content and 350-450 $kg/m^2.h$ specific filter output. The high capital and operating costs and the complicated engineering and control have so far militated against use of this filter under practical conditions.

Continuous pressure filtration and combined pressure/vacuum filtration for the dewatering of coal slurries were investigated in a drum filter pilot plant at the Institute for Mechanical Process Engineering and Mechanics at Karlsruhe Technological University (22). Although a material which was particularly difficult to dewater was used with the concentrate of a pneumatic flotation with an ultra-fine proportion below 0.063 mm of about 65% by weight, such a good result was achieved with regard to the residual water content and specific filter output that practical tests are now being prepared for continuous pressure filtration of coal slurries in two coal preparation plants.

4.2.3 Special filters

Special filters used in other branches of industry for dewatering ultra-fine solids have so far not been tested in coal preparation, because their operation involves extremely high costs and the mass flows to be processed in the mining industry are too large with the present stage of development of these special filters. However, they may be of interest in the course of the more recent developments, in which relatively small quantities of ultra-fine solids are to be separated.

5. CHEMICAL DEWATERING AIDS

It has been ascertained in comprehensive laboratory tests that the dewatering of fines and ultra-fines can be significantly improved by addition of surfactants (9,23). The reduction of the water content was confirmed in initial tests; however, attendant phenomena such as excess frothing and an adverse effect on flotation occurred, so that further basic investigations are required before use of chemical dewatering aids under practical conditions.

6. CONCLUSIONS

Improvement of dewatering of fines and ultra-fines is and will remain a main area of research and development in the coal preparation sector. Whereas essentially the changes in raw material conditions were the determinative factors in the past, it is now the market requirements for low-sulphur high-grade coal, which necessitate intensified grading in the ultra-fine range.

In addition to improvement and further development of known and the testing of new dewatering equipment not previously used in coal preparation, the subdivision of the particle range under about 10 mm into more than the hitherto customary two sub-ranges and the separate dewatering of these narrow sub-ranged in specially designed machines is apparently one way of observing water contents in the fine and ultra-fine products and limiting the application of thermal drying to an unavoidable minimum even under difficult conditions.

A series of individual investigations, some of which were quite successful, have been conducted in the past. These must be continued, whereby the solids discharge of the centrifuges must be increased and the residual water content of the filters reduced.

The final aim of these considerations and this work must be to revise the concept of dewatering in the full particle size range under about 10 mm, which also takes into account in particular the ultra-fine residual solids discharged in the concentrates and filtrates.

7. BIBLIOGRAPHY

1. Kundel, H: Face technology in the German coal mining industry in 1981. Glückauf 118 (1982), pp. 938-947.

2. Lemke, K: Technical and economic effects of the deterioration of raw coal properties. Glückauf 96 (1960), pp. 1257-1263.

3. Bartelt, D: Effects of the new maximum workplace concentrations on moisture of the raw coal and thus on the economic result of the mining industry. Appendix 3 to the minutes of the 8th meeting of Committee D at the Steinkohlenbergverein on 27.2.1972.

4. Bethe, W.P: Future requirements on coal preparation by raw coal and market developments. Glückauf 116 (1980), pp. 1117-1123.

5. Michalowski, B: Investigations into the dewatering of washed fine coal in a tumbler centrifuge. Aufbereitungs-Technik 23 (1982), No. 6, pp. 302-305.

6. Padberg, W: Final report on the development project "Basic research for mechanical subsequant dewatering of ultra-fines in the centrifugal field". Saarbrücken, 1981.

7. Padberg, W: Final report on the development project "Mechanical dewatering in the ultra-fine range". Saarbrücken, 1978.

8. Bogenschneider, B: et al Dewatering of coal flotation concentrate particularly with solid bowl screen centrifuges. Glückauf 116 (1980), pp. 1006-1012.

9. Erdmann, W: Final report on the research project "Optimisation of the dewatering of fine preparation products". Essen 1983.

10. Wilczynski, P: Final report on the development project "Solid bowl centrifuge for the dewatering of flotation tailings". Essen 1977.

11. Legner, K. and J. Schwerdtfeger: Dewatering of ultra-fine coal slurries on a drum filter with belt removal. Glückauf 119 (1983), pp. 237-239.

12. Hennig, H. et al: Initial use of vacuum belt filters and filter belt presses as dewatering machines for ultra-fines in the coal mining industry at the Lohberg mine. Aufbereitungs-Technik 20 (1979), No. 4, pp. 177-191.

13. Blankmeister, W: Results of the investigations of a vacuum belt filter for dewatering coal concentrate. Aufbereitungs-Technik 21 (1980), No. 4, pp. 171-176.

14. Kubitza, K.-H. and H. Lüdke: Final report of the Steinkohlenbergbauverein on the development project "Optimisation of the properties of the fine preparation products" (partial project: improvement of vacuum filtering). Essen 1976.

15. Koppitz, K. and H. Kozianka: Final report on the research project of the Federal Minister for Economic Affairs, project No. 594004 "Steam hood filter for dewatering of flotation concentrate".

16. Jones, G. et al: Optimisation of the operation of a membrane filter press. Contribution J.4 to the 9th International Congress on Coal Preparation. Delhi, India, 29.11.-4.12.1982.

17. Singh, B.K. and W. Erdmann: Dewatering of flotation tailings in screen belt presses. Glückauf 114 (1978), pp. 305-309.

18. Erdmann, W: Final report on the development project "Optimisation of the preparation of washing water and other suspensions". Essen 1981.

19. Erdmann, W. and K.H. März Final report on the development project "Comparative operation of filter belt presses, solid bowl centrifuges and chamber filter presses for dewatering flotation tailings under comparable raw material conditions with special consideration of the further development of filter belt presses". Essen 1983.

20. Dosoudil, M: Continuous pressure filters. Chemie-Technik 8 (1979), pp. 397-399.

21. Dosoudil, M: Application of the continuous pressure filter to flotation concentrates. Aufbereitungs-Technik 25 (1984) No. 5, pp. 259-265.

22. Bott, R. et al: Continuous pressure filtration of ultra-fine coal concentrates. Aufbereitungs-Technik 25 (1984), No. 5, pp. 245-258.

23. Hoberg, H. et al: Improvement of the dewatering of ultra-fine coal. Paper given at the 20th conference on "Mechanical liquid separation", Magdeburg, 3 and 4 November, 1983.

Figure 1. Jigging screen centrifuge with 1500 mm basket diameter.

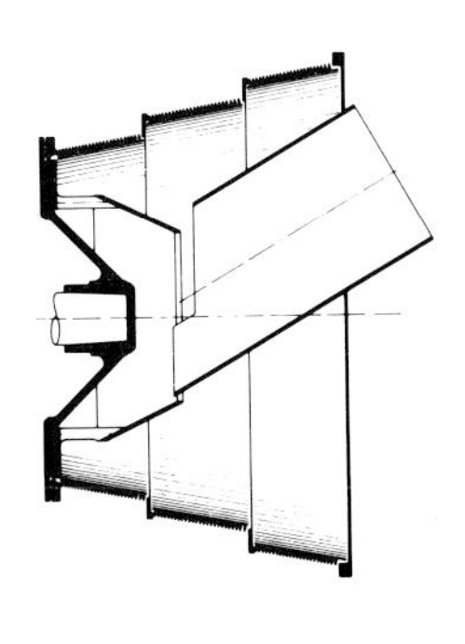

Figure 2. Stepped basket for jigging screen centrifuge.

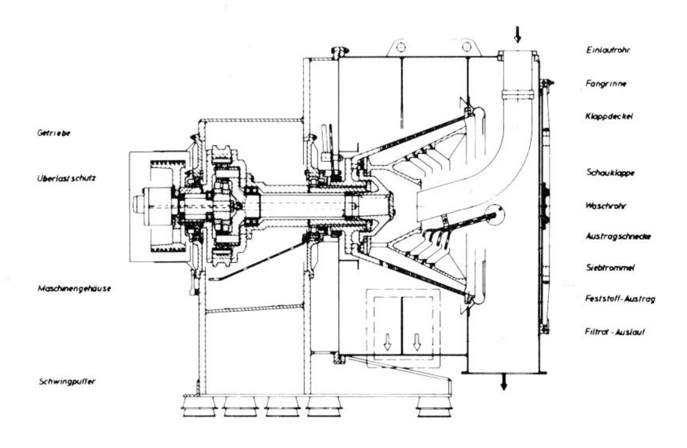

Gebriebe	gear
Überlastschutz	overload protection
Maschinengehäuse	machine casing
Schwingpuffer	vibration dampers
Einlaufrohr	feed pipe
Fangrinne	collecting channel
Klappdeckel	hinged cover
Schauklappe	inspection flap
Waschrohr	washing pipe
Austragschnecke	discharge screw
Siebtrommel	screen drum
Feststoff-Austrag	solids discharge
Filter-Auslauf	filter outlet

Figure 3. Horizontal screw screen centrifuge.

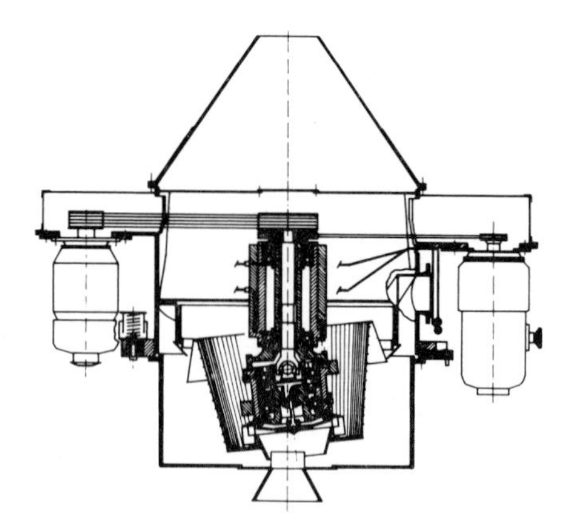

Figure 4. Tumbler centrifuge.

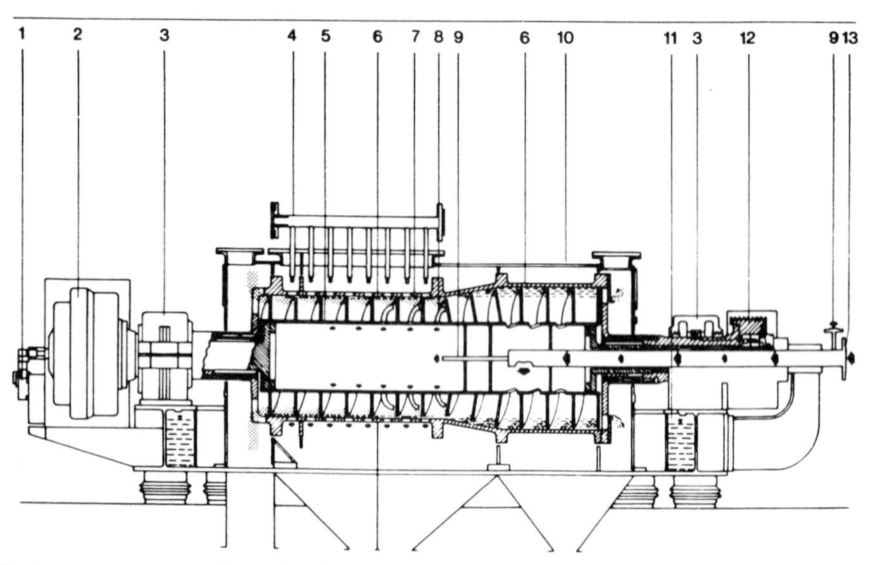

1 torque measuring device
2 planetary gear
3 pedestal bearing
4 screen washer
5 screw
6 drum with cylindrical screen section
7 fine screen covering
8 washing pipes
9 washing water
10 casing
11 oil bath for circulation lubrication
12 drive pulley
13 feed pipe

Figure 5. Solid bowl screen centrifuge.

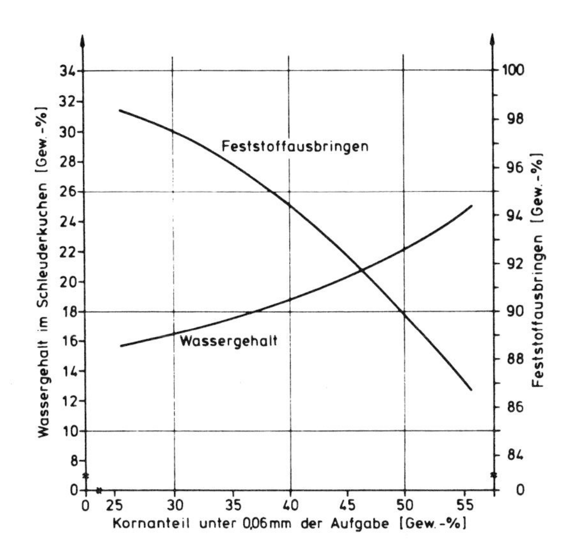

Wassergehalt im Schleuder-kuchen (Gew. - %)	water content in centrifuge cake (% by wt.)
Kornanteil unter 0.06 mm der Aufgabe (Gew. - %)	feed fraction below 0.06 mm (% by wt.)
Feststoffausbringen (Gew. - %)	solids yield (% by wt.)
Feststoffausbringen	solids yield
Wassergehalt	water content

Figure 6. Dewatering result of a solid bowl screen centrifuge.

MEMBRANE PRESSURE PLATE FILTRATION

G S JONES

COAL PREPARATION DIVISION, MINING RESEARCH AND DEVELOPMENT ESTABLISHMENT,
NATIONAL COAL BOARD

Summary

It has been shown conclusively that specific advantages exist for the production of drier filter cakes with increased throughput when using filter presses equipped with membrane plates.

The potential of a membrane plate filter press has been further examined, with particular reference to optimisation and automatic control. Practical trials have been carried out on a production installation.

Due to unavoidable delays only limited results are available but there is an indication that, eventually, automatic control of a membrane plate filter press could be derived from pre-set values of cake thickness and moisture content.

In the short term, the continuing use of membrane plate filter presses must be in some doubt following the total failure of proprietary 2 m x 1.5 m membrane plates at two National Coal Board filter press installations. Alternative plate designs are under active consideration in an attempt to re-establish the reliability demonstrated by earlier trial installations using plates of dimension 1.3 m x 1.3 m.

1. INTRODUCTION

The large quantities of waste material presented for disposal from coal preparation plants creates enormous problems, in particular the water-borne fraction below 0.5 mm. Between 4 and 6 million tonnes of material of this size fraction has to be disposed of each year, the bulk of which being classed as flotation tailings.

The modern large-scale coal preparation plant produces flotation tailings at rates of between 50-180 tonnes/hour (dry solids). A two-product separation is required to return the bulk of the water (from which the solid particles have been borne) back into the washing circuit and to produce solids in a form which allows safe disposal onto the dirt stack. Lagooning is not considered acceptable because of environmental reasons. Of the different mechanical techniques that are available, pressure plate filters, although high in capital and operating costs and relatively low in output per unit machine area, are still considered to be one of the most satisfactory.

The majority of large-scale filter presses used by the National Coal Board are 2 m x 1.5 m x 100-150 chambers, producing a filter cake nominally 32 mm thick. The total dry weight of filter cake produced each cycle is approximately 9 tonnes and, depending on the amount of very fine material present, especially in the region of 10 μm, a typical total time for a complete filter press cycle can vary from 1 hour to 4 hours. The end moisture content of the filter cake is critical, although it is not uncommon for operators to extend the filter press cycle even further to ensure a cake that can be discharged cleanly from the press.

Initial trials using a small trailer-mounted pilot filter press fitted with membrane plates and subsequent larger scale trials using membrane plates fitted to 100-chamber mechanised filter presses at specific collieries (1), indicated that press cycle times could be substantially reduced. The output rate of filter cake was thereby increased. In addition, it was found that the cake moisture content was also significantly reduced.

Based on this evidence, it was decided to further examine and develop the application of membrane plates in the filtration of tailings and effluents.

2. MEMBRANE PLATE FILTER PRESSES

A typical UK membrane plate construction consists of an internal body of mild steel and ebonite, around which is moulded a rubber membrane to form a complete envelope. The outer recess faces of the membrane are moulded with a drainage pattern of 'pips' to provide adequate drainage of filtrate behind the backing cloth. An air inlet is provided through the steel insert so that compressed air can be applied between the internal body and the rubber membrane, to inflate the membrane and squeeze the cake contained in the recess. Filtrate is removed from the outer surface of the membrane through drainage tubes between the gasket faces of the plates. These membrane plates are normally used alternately with conventional rubber-coated plates to form a press pack in which each chamber of the press has one fixed face and one flexible face.

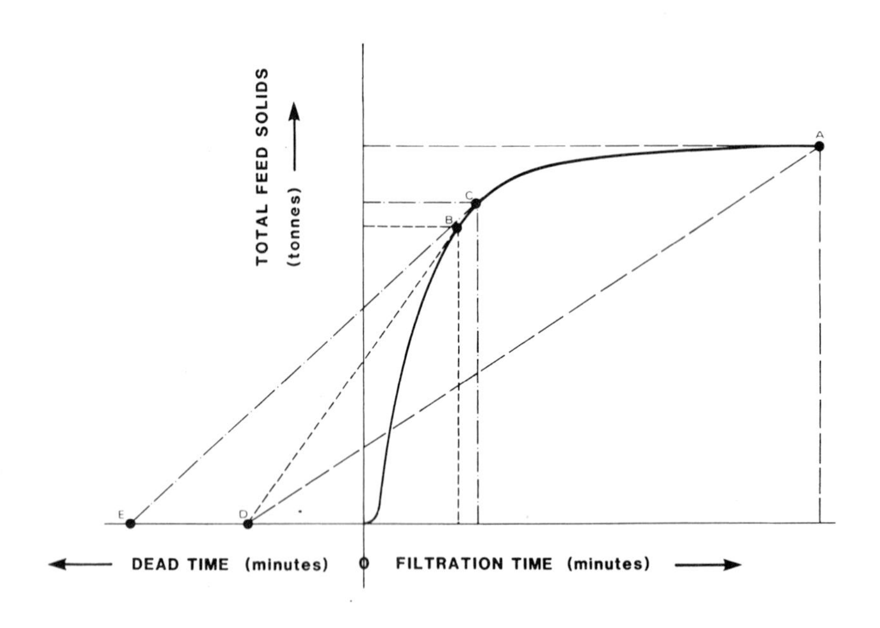

Figure 1 Optimisation of a Filter Press

A diagram to explain the potential of the membrane plate press is shown in Figure 1.

With a conventional press of fixed volume it is necessary to fill the plate recesses with a cake of sufficiently low moisture content to discharge easily and to be stable on a dirt stack. This necessitates the operation of the press for a sufficient time to reach the cycle termination point A. The time taken to achieve this position is the filtration time plus a further period of 'dead' time for opening, discharging and closing the press ready for the next cycle. The overall throughput rate is represented by the slope of the line DA. If it were possible to terminate the pressing at a point B in a very much reduced filtration time, the throughput rate would be substantially increased, represented by the slope of the line DB. However, because the conventional press is of constant volume this increased throughput rate is achieved at the expense of a much wetter cake which would be impossible to discharge cleanly and would not be acceptable for dirt stack disposal.

With a membrane plate press, however, the pressing can be terminated at any stage and the membranes inflated to express water from the cake. Since this part of the press cycle does not introduce additional solids into the press, it can be regarded as an extension of the dead time. The dead time is extended, therefore, to point E. A tangent from point E to the throughput curve gives the maximum overall throughput and the required time for pressing. This occurs at point C, the slope of the line EC representing the maximum throughput rate. Conditions are created therefore where a filter cake of the same moisture content as that achieved at point A by conventional pressing, is achieved at point C, together with a substantial increase in throughput rate.

This potential has been the subject of practical trials (1) and definite trends in performance have been observed. It has been shown that, at the lower moisture contents achieved by conventional pressing, there is a substantial increase in throughput if the same moisture contents are produced using the membrane cycle. (Shown by using actual filter press cycle values, to be in excess of 40%).

A further advantage of the membrane plate press is that when required, the moisture content of the cake can be reduced to a level which a conventional press, operated at the same pressure, cannot achieve. Direct application of force over the whole area of the cake is much more efficient than hydraulic pressure applied by the feed fluid through the eye of the press.

3. OPTIMISATION

Even though modern coal preparation plants are computer controlled, the termination of a filter press cycle is still left to the judgement of an operator. The termination point is normally governed by the rate of filtrate discharge or an empirical set value of time for the pressing cycle. Neither system adequately takes into consideration changes in the filtration characteristics that can occur quite rapidly due to changes in the quality of the flotation tailings being fed to the press. With a membrane plate press there is also the additional requirement of knowing when to stop hydraulic filtration and introduce the membrane squeeze.

Filter cake moisture content is important in terms of how easy cake is discharged from the press and also its handlability and eventual disposal onto the dirt stack. The filter cake throughput rate of the press is also of prime importance.

3.1 Automatic Optimisation of a Variable Volume Filter Press to give Maximum Throughput Rate

The ability of a variable volume (membrane plate) filter press to work at the optimum cake thickness to give maximum throughput rate is fairly well known but the selection of the optimum cake thickness requires considerable work even when the filtration characteristics of the feed slurry remain constant. When the characteristics of the slurry are variable the selection of optimum cake thickness by normal manual methods is almost impossible.

Automatic optimisation requires transducers to continually measure the rate of feed flow to the press and the solids content of the feed, and a microcomputer to calculate the solids flow rate.

When the throughput rate of the press has reached a maximum the hydraulic pressing stage of the cycle is terminated. The membrane squeeze is then applied.

The hydraulic pressing stage is the period when filtration takes place, ie from the moment the press feed pump(s) are started to the start of the membrane squeeze. The membrane squeeze, incorporated in the "dead time" period, is added to the hydraulic pressing stage to give a measurement of total running time.

Solids throughput rate is computed from:

$$\frac{\text{TOTAL MASS OF FEED - TOTAL MASS OF LIQUID}}{\text{TOTAL TIME}}$$

This equation is expressed in the form:

$$\frac{\frac{\rho s}{\rho s-1}\left[\int Q\rho_f \, dt - \int Q dt\right]}{Dn + t}$$

where ρs = Relative density of the solids

ρf = Relative density of the feed slurry

Q = Volumetric flow of feed slurry to the press

Dn = Dead time

t = Hydraulic filtration time.

3.2 Automatic Termination of a Filter Press Cycle According to Moisture Content of the Filter Cake

Because cake moisture content is an important factor in terms of cake release and disposal it is of advantage to be able to determine the moisture content in situ and to terminate the press cycle when the moisture content has been reduced to a required value.

Moisture content of the material within the press is computed from:

$$\frac{\text{TOTAL FLUID INPUT - TOTAL SOLIDS INPUT - TOTAL FILTRATE}}{\text{TOTAL FLUID INPUT - TOTAL FILTRATE}}$$

This equation is expressed in the form:

$$\left(\frac{\frac{1}{\rho s-1}\left[\rho s \int Q dt - \int Q\rho_f \, dt\right] - \int F dt}{\int Q\rho_f \, dt - \int F dt}\right) \times 100$$

All the computations are carried out continuously and the calculated moisture content compared with the required value.

When the required value is reached, the press cycle is automatically terminated.

4. ECSC AIDED PROJECT 7220-EA/807

Initial trial installations of the membrane plate filter presses showed definite trends in improved performance, related to both cake moisture content and solids throughput. These trials were carried out on single 1.3 m x 1.3 m x 100 chamber filter presses located at Littleton and Elsecar Collieries (1).

Encouraging results from both trial installations resulted in all the presses being converted to membrane plate operation, 5 presses at Littleton and 3 presses at Elsecar.

Filter presses are used throughout the European Mining Community. For a technically satisfactory process the continuing aim is to increase throughput and reduce costs. Project 7220-EA/807 was initiated to carry out the following:

(i) Prove that improved performance, as indicated by the initial membrane plate filter press trials, can be maintained over extended periods of operation in a continuous production stream.

(ii) Examine the mechanical aspects of the system with a view to increased reliability.

(iii) To optimise the process performance and progress towards automatic control of the total system using a microcomputer.

Since all the filter presses at both Littleton and Elsecar Collieries were now converted to membrane plate operation, objectives (i) and (ii) could easily be carried out by long-term observation of the normal operation of the filter cake production process.

To achieve objective (iii) it was decided to convert one of the Elsecar filter presses to a fully optimised master control press with the hope that eventually the other two presses could be slave driven thus achieving total system control.

4.1 Objective (i) - Membrane Filter Press Performance

Trial sites at Elsecar and Littleton Collieries have satisfied the project objective (i). At Littleton Colliery, membrane plate filter presses have been in use since 1980, the improved performance resulting in recoverable proceeds, ie 20% of previously discarded filter cake now being used as a saleable product. At Elsecar Colliery, improved performance resulted in the total plant capacity being maintained when a rapid deterioration in the quality of the flotation tailings (high slimes content) occurred. Continuous filter cake production, using all the converted presses, ran from March 1981 to November 1982.

4.2 Objective (ii) - Membrane Filter Press Reliability

Other than difficulties associated with converting very old presses at Littleton Colliery there have been no serious reliability problems in the use of 1.3 m x 1.3 m membrane plate filter presses over a period of operation as outlined above.

The success at both of these trial sites encouraged the National Coal Board to install 3 new membrane plate filter press installations, using a total of 25 - 2 m x 1.5 m x 100 to 150 chamber proprietary machines. At two of the sites, commissioned in February and November 1982, problems of inferior design and construction led to a total failure of the membrane filter press plates. The filter presses have yet to be commissioned at the third site, although they consist of plates of a similar design to those used at the other site trials. Whether or not these plates fail in service has yet to be determined. The failures have obviously been a set-back for the development of membrane plate filtration and remedies are urgently being examined.

4.3 Objective (iii) - Membrane Filter Press Optimisation

The conversion of one of the Elsecar filter presses was carried out in conjunction with the requirements as described in Section 3.

A microcomputer was required which would sequentially control the operation of the filtration cycle, perform calculations from data submitted by measurement transducers, issue instructions to control the filtration process, and communicate with a data logging system.

The data logging system was required to process and store extensive data collected from a filtration cycle of 2-3 hours duration, to display recorded data in both graphical and tabular form and to be portable for off-site data examination.

The eventual choice of microcomputer was a TEXAS PM 550 system. For data logging, a HEWLETT PACKARD HP85 desk-top computer was selected which included an in-built magnetic tape storage facility, visual display unit and printer.

Measurement transducers initially chosen were magnetic flow meters for the feed and filtrate flow rates and a nucleonic density gauge for the density of the feed pulp.

To complete the package, a mimic process flow diagram was provided where all measured values were displayed, and a key-pad incorporated to input fixed data and filtration cycle termination points.

The total package has the facility, therefore, to provide full automatic control of the filter press, to record all input and output data, to carry out continuous calculations of variables, such as solids throughput and cake moisture content, and to scan and record 14 separate items of dynamic data every 8 seconds over the whole of the filtration cycle (ranging between 0.5 and 4 hours).

The above package was incorporated into the main filter press plant control panel and interlocked in such a way that all three filter presses could eventually be driven using optimised data for filtration cycle termination.

4.3.1 Optimisation trials

Software for the optimisation control system was prepared to terminate the input of feed slurry at the point of maximum solids throughput rate and to terminate the filtration cycle according to a pre-set moisture content of the filter cake.

Fixed inputs to the microcomputer included the desired final cake moisture content, Dn constants, solids relative density constants and solids ash content of the filter press feed material.

Commissioning of the optimisation control system, based on the above termination points, revealed a number of basic problems including communication between the microcomputer and the data logger and negative cake moisture values. These problems were eventually overcome.

An error was discovered in the filtrate flowmeter output due to air entrainment in the filtrate measurement circuit. This was resolved by repositioning the magnetic flowmeter. Excessive errors were also discovered in the output from the feed slurry magnetic flowmeter.

A further major problem was the point at which the maximum solids throughput to the filter press was reached. Optimisation resulted in very thin cakes, too thin for easy removal from the plate recesses. The solution to this problem was to update the software so that the termination of the input of solids to the filter press could be at a pre-set cake thickness. This cake thickness would be based upon (a) the ease of discharge from the press and (b) in the knowledge that a chosen thickness less than the recess depth dimension would still give an improvement in throughput capacity.

Further proving trials using the updated software continually revealed errors of actual cake thickness and moisture contents to the pre-set values. Transducer errors were still considered to be the main problem. Detailed investigation of the feed flow magnetic flowmeter system revealed that the ram pump peak flow rates were not being recorded due to the slow response time of the flowmeter electronics. At this point it was decided to abandon the magnetic flowmeter as the measurement transducer of feed flow to the filter press and to replace it with a system incorporating load cells to measure the weight of slurry being fed to the press.

The filtrate flow measurement system was slightly modified by increasing the velocity of filtrate through the magnetic flow meter and introducing dual range electronics.

The nucleonic density gauge used for measuring the pulp relative density was also checked for accuracy.

After such exhaustive investigations the transducer errors were reduced to the following:

Feed input (load cells) ±0.02%
(previously with magnetic flowmeter 7% average error recorded)
Filtrate flow ±1% of total flow
Pulp relative density ±0.01% (absolute)

One remaining source of error was the solids ash content figure used to derive the solids relative density. As yet there is no proprietary device available to instantly measure the ash content of a tailings slurry. The current method is to use a previously calculated filter cake solids ash value derived from a feed or cake sample at least 3 h prior to the values entry into the microcomputer. With varying tailings solids ash values it is not possible to accurately predict the instantaneous ash value. (The derivation of solids RD is related to a reciprocal of the actual ash value.)

Currently, a joint MRDE/AERE (Harwell) project is being pursued to develop an on-line tailings ash monitor. One of the first prototypes has already been installed for trials before eventual inclusion into the optimisation control circuit.

Having reached a specific stage in the commissioning of the optimisation control system it has not been possible to carry out a complete trials programme as originally planned. The closure of Elsecar Colliery was announced in March 1983 and considerable delay occurred in locating a further site and installing an identical system.

This was eventually achieved in January 1984 at Orgreave Coal Preparation Plant, South Yorkshire Area.

At the point when a full programme of trials had been scheduled, the National Union of Mineworkers commenced industrial action and no further work at that site has been possible.

The results already obtained, therefore, in terms of objective (iii) only offer guidelines as to the future possibilities of membrane plate pressure filtration.

4.3.2 Results

Optimisation trials were carried out using flotation tailings as the feed to the filter presses. Filtration characteristics, however, varied considerably with a filtration cycle time at Elsecar averaging 2 h and at Orgreave approximately 40 min. The reason for this difference was the high slimes content of the shales at Elsecar indicated by 80-85% of material below 63 μm and Orgreave producing less than 60% of material below 63 μm.

The set points for cake thickness and moisture ranged from 20-32 mm and 19-32% w/w respectively at Elsecar and 25-27 mm and 20-24% w/w at Orgreave.

Table I summarises the data analyses of all the useful filtration cycles performed at both sites. Results are displayed to show the number and percentage of filter press cycles occurring at less than 1, 2 and 3 mm deviation from the cake thickness set point and 1, 2 and 3% absolute moisture deviation from the cake moisture set point. The number of cycles outside these limits are also shown.

ELSECAR COLLIERY				
DEVIATION FROM SET POINT		NUMBER OF PRESS CYCLES (% OF TOTAL)		
FILTER CAKE		FILTER CAKE		
THICKNESS (t) mm	MOISTURE (m) % w/w (absolute)	THICKNESS (t)	MOISTURE (m)	COMBINED (t) & (m)
<1	<1	24 (32.9)	10 (13.7)	4 (5.5)
<2	<2	35 (47.9)	27 (37.0)	12 (16.4)
<3	<3	50 (68.5)	37 (50.7)	25 (34.3)
>3	>3	23 (31.5)	36 (49.3)	48 (65.7)
TOTAL NUMBER OF PRESS CYCLES				73 (100)

ORGREAVE COLLIERY				
DEVIATION FROM SET POINT		NUMBER OF PRESS CYCLES (% OF TOTAL)		
FILTER CAKE		FILTER CAKE		
THICKNESS (t) mm	MOISTURE (m) % w/w (absolute)	THICKNESS (t)	MOISTURE (m)	COMBINED (t) & (m)
<1	<1	4 (44.4)	4 (44.4)	1 (11.1)
<2	<2	6 (66.7)	6 (66.7)	4 (44.4)
<3	<3	8 (88.9)	7 (77.8)	6 (66.7)
>3	>3	1 (11.1)	2 (22.2)	3 (33.3)
TOTAL NUMBER OF PRESS CYCLES				9 (100)

TABLE I : SUMMARY OF ELSECAR & ORGREAVE RESULTS

Filter press cycles at Elsecar Colliery, although of a significant number (73), must be viewed in the light of transducer errors (as explained in section 4.3.1). Orgreave results, however, show a definite improvement, and due in part to control system modifications. The results can only be viewed, however, in the context of the limited data analysis (9 filtration cycles).

5. CONCLUSIONS

It has been shown quite conclusively, both in other industries and the National Coal Board (1), that specific advantages exist for the production of drier filter cake and increased throughput when using filter presses equipped with membrane plates.

It has also been shown theoretically that by using a membrane plate press it is possible to determine the optimum cake thickness and to calculate the moisture content of the filter cake in situ within the press. Termination of the press cycle occurs at the point of optimum cake thickness and when the moisture content has a pre-set value.

This theory has been put into practice with limited success. Certain factors exist, however, that prevent a completion of the objective to achieve full automatic optimisation control of a filter press cycle. These factors are as follows:

(i) The maximum throughput rate is reached very quickly in the filtration cycle and if the press operation is terminated at this point, thin cakes would have to be discharged. Since it is often difficult to discharge standard 32 mm thick cake from existing filter presses, total automatic discharge of thinner cakes cannot be expected. The development of existing filter presses must, therefore, be accelerated to adapt to the opportunities that exist for total control.

(ii) Regardless of the difficulties as outlined above, investigations continued on the basis of selecting any desired cake thickness, albeit at less than 100% optimisation, together with a pre-set cake moisture for press cycle termination.

Reference to Table I, however, shows how limited the success of the investigations has been. The small proportion of results falling within acceptable deviation limits of the set points can almost certainly be due to individual transducer error, some of which were considerable during the early stages of the investigation. Fortunately, these have now been reduced to a more acceptable level, the more recent Orgreave results indicating an improvement, although many more tests are required to confirm this trend.

A continuous and accurate signal of tailings ash is the one remaining variable that has not been improved. It is hoped that the tailings ash monitor currently being developed jointly by the National Coal Board and the Atomic Energy Research Establishment will provide this answer.

(iii) The closure of Elsecar Colliery, the delay in finding an alternative site (Orgreave) and the current industrial action taken by the National Union of Mineworkers has curtailed an original 3 year development project by 1.5 years. Had normal operation been possible the improvement observed at Orgreave could have been confirmed by many more operational cycles. When possible, investigations will continue.

The continuation of membrane plate filter press development is also dependent on the availability of a reliable and functional membrane plate. The failure of the 2 m x 1.5 m membrane plates at two large NCB filter press installations dealt a damaging blow to this type of technology. Fortunately there are alternative designs of membrane plate available from other manufacturers and these are currently under active consideration.

6. ACKNOWLEDGEMENTS

The author wishes to thank Mr T L Carr, Head of Mining Research and Development for permission to publish this paper.

Full appreciation is also given to the European Coal and Steel Community under whose sponsorship this work was carried out. The views expressed are those of the author and do not necessarily represent those of the National Coal Board.

7. REFERENCES

1. JONES, G S, et al "The Operation and Ultimate Optimisation of a Membrane Plate Filter Press". (Paper presented to the Ninth International Coal Preparation Congress, India, November 1982).

THE INFLUENCE OF SURFACE-ACTIVE AGENTS IN WASHING WATER ON FLOTATION AND FLOCCULATION

W. Hey
Scientific Staff Member
Bergbau-Forschung GmbH
Preparation Department

Summary

The selective separation of coal and dirt in bituminous coal flotation crucially depends on the specific effect of a flotation agent on the wettability of solids particles and thus on the interfacial effects between solids surface and liquid surface. When using reagents as filtering and sedimentation aids this proposition is broadly valid for filtration and flocculation. By means of model tests carried out in the laboratory, in which several substances were used differing in their chemical composition - it proved possible to quantify the main factors influencing the wetting and adsorption process. Research concentrated on the physical-chemical properties of the solid/liquid interface. The significance of individual measured values were estimated by correlation of investigation results. Surfactant model substances and technical flotation agents were available for the examination of capillary-active properties of flotation additives and on the influence of these substances on the wetting, adsorption and flotation behaviour of the preparation product. Also included in the tests were a non-ionogenous surfactant and a variety of flocculants.

1. INTRODUCTION

Flotation is a complex and not very transparent macro-process in which a multitude of chemical, physical and physico-chemical micro-processes take place at phase interfaces partly overlapping, partly in sequence. The adhesion of the flotation particles to air bubbles is crucial for the effectiveness of flotation (1). Decisive for adhesiveness to an air bubble is the contact angle which the air bubble forms with the solids surface once adhesion has occurred (2). The contact angle and the induction time are influenced by the degree of coalification and the surface properties of the coals used, on the pH value of the solution, its ion concentration, the Redox potential with regard to the coal, the chemical composition of the collector and not last by surfactants and organic polyelectrolytes as water carriers (3-9). From this it emerges that flotation is a process with many-layered interactive forces, and the question arises as to the real physical and physico-chemical causes of the selectivity and the interaction of different reagents. Examination of the effect of the adsorption layer on the flotation-relevant processes is something which is undoubtedly in the forefront here, the difficulties primarily being the fact that it is not yet possible to apportion unmistakable physical and chemical parameters.

In coal preparation the flocculants used in fines thickening and wash-water clarification, in filtering and partly in dewatering in centrifuges are based on polyacrylate and polyacrylamide. Since the water is mainly kept in a circuit within the preparation plant after clarification, i.e. it is re-used, flotation with flocculants is in a sense affected in advance. In practice, reciprocal influences between flotation and sedimentation reagents have often been established as causes of breakdowns, and it is this that inspired the investigations which are now described.

2. CHARACTERISATION OF TEST SUBSTANCES

A gas coal and dirt were used as solids for the investigations for determining wetting head, adsorption isotherms, sedimentation behaviour and flotation selectivity. The upper size limit of the particle package was 500 µm. Screening produced four particle size fractions - 500-315 µm/315-125µm/125-63 µm/ below 63 µm - so that the parameter of particle size could be brought into the investigations. Particle size distribution and the relevant solids data are collated in Table I.

		Size Proportion	Ash % (wf)	V.M. % (waf)
Gascoal	$<$ 500 µm	100.00	3.69	33.92
Fractions	500-315 µm	4.29	4.76	
	315-125 µm	26.32	3.08	
	125- 63 µm	20.81	2.64	
	63 µm	48.58	4.90	
Dirt	500 µm	100.00	93.20	

Table I: Solids data

For the flotation investigations a model mixture made up of gas coal and dirt was used with a particle size proportion of 52.7% below 63 µm and an ash content of 35%.

The flotation reagents used in coal flotation to make solids surfaces water-repellant, available as ready-made commercial products, are in effect mixtures of higher aliphatic alcohols with a chain length of C_6 - C_{10}. In most cases they are collector-frothing agent combinations. Four different commercial brands were examined, as was the typical flotation alcohol 2-ethylhexanol-(1) present in all mixtures. Investigations also covered three flocculants differing in iorigenity and molecular weight and a non-ionogenous wetting agent of the polyglycolether type.

Flotation agents: Montanol 350
Montanol 551
Ekofol 40/25
Ekofol 452
2 ethylhexanol-(1)

Flocculants: P 2810 (anionic)
P 2935 (anionic)
P 6000 (non-ionogenic)
P 523 K (cationic)

Surfactants: Eumulgin 286 (non-ionogenic)

3. WETTING HEAT AND ADSORPTION ISOTHERMS

As adsorption processes are generally exothermic, it is possible to obtain evidence on the type of interactive forces by means of thermal measuring methods. The amount of heat developed during the wetting of a solid interface by a liquid is termed wetting heat. This wetting heat, being specifically related to surface, is basically more accessible than wetting tension, because it can be thermally measured directly by introduction of the solid into the liquid. Since the surface is generally not known with very fine solids, one relates to 1 g of solid substance in each case instead of a surface unit, something which is often sufficient because the comparisons of different liquids always relate to the same solid.

The practical procedure starts by stirring a predetermined sample into a liquid. The wetting heat can then be calculated from the resulting temperature rise and the specific heat of the liquid.

Table II shows the measured wetting and reaction heats of different water carrying substances compared with coal and as between the solutions. The wetting heat of water, methanol and benzene in contact with coal illustrated the water-resistant character of coal. The wetting heat of water is considerably lower than that of methanol and benzene. The lower figure of benzene can be attributed to the steric impedence to surface covering, especially in the micro-pore range. If a 0.1% polymer solution is mixed into the water content of the calorimeter a negative mixing heat is obtained for all three polymerates. But if the polymer solutions are mixed into a water-coal suspension negative heat values are also obtained but these are considerably lower than the values compared with pure water. The difference between the polymer-water values and the polymer-coal-suspension values can be regarded as the adsorption heat of polymers with coal.

Calorimeter content	Admixture	Result		
		W_B (J/g)	W_M (J/g)	W_A (J/g)
100 ml water	10 g coal	8.8		
100 ml methanol	10 g coal	21.6		
100 ml benzene	10 g coal	14.7		
100 ml water	10 ml sol. 2810		-4.4	
100 ml water	10 ml sol. 6000		-4.3	
100 ml water	10 ml sol. 523 K		-4.4	
100 ml water + 10 g coal	10 ml sol. 2810		-0.59	3.8
100 ml water + 10 g coal	10 ml sol. 6000		-2.05	2.3
100 ml water + 10 g coal	10 ml sol. 523 K		-0.90	3.5
100 ml Montanol sol. (0.01%)	10 g coal	7.1		
100 ml Montanol sol.	10 ml sol. 2810		-3.3	1.1
100 ml Montanol sol.	10 ml sol. 6000		-2.9	1.4
100 ml Montanol sol.	10 ml sol. 523 K		-2.9	1.5

TABLE II: Wetting head of different additives. (Coal particles size $< 63\ \mu m$, W_B = wetting head, W_M = mixing heat, W_A = adsorption heat of released molecules).

The admixture of polymer solutions to a Montanol solution shows a lesser negative heat gradation than with an admixture of polymers to water. Therefore, an interaction between Montanol and polymer is to be assumed.

The adsorption of flotation and flocculant molecules in water to coal or dirt particles is primarily a physical adsorption, i.e. the binding to the solids surface is characterised by Waals forces. In adsorption an equilibrium sets in sooner or later which depends on the prevailing conditions. It can be generally described by a function $f(Q, c, T) = 0$, where Q is the quantity (g or mol) of the adsorbed material per mass m of the adsorbent, c the concentration of the material to be adsorbed and T the thermodynamic temperature. When measuring adsorption equilibria T is kept constant to obtain the so-called adsorption isotherms $(Q)_T = f(c)$.

The following adsorption isotherms of Eumulgin (surfactant) and ethylhexanol (flotation alcohol) show in what way the two components influence one another and what effect organic polymers have on adsorption. The results of adsorption isotherms of Eumulgin and ethylhexanol to coal and dirt are recapitulated in Table III. Each isotherm consists of four measurement points.

Adsorbent	Adsorptive	Residual concentration C_R (mg/l)	Charging Q (mg/kg)
Gas coal (500 - 0 μm)	Eumulgin	30.4	1 806
		10.9	783
		4.0	388
		2.0	195
Dirt (500 - 0 μm)	Eumulgin	15.1	1 916
		4.3	948
		0.7	411
		0.3	190
Gas coal (500 - 0 μm)	Ethylhexanol	26.4	1 485
		9.0	802
		1.4	358
		0.3	188
Dirt (500 - 0 μm)	Ethylhexanol	87.1	246
		40.4	159
		16.4	69
		7.2	48

TABLE III: Adsorption isotherms of Eumulgin and ethylhexanol to coal and dirt

Table III shows that the Eumulgin surfactant adsorbs very well to both coal and dirt, while ethylhexanol exhibits only slight adsorption to dirt. The adsorption to coal of Eumulgin is also higher than that of ethylhexanol. Eumulgin, therefore, adsorbs well independently of the adsorbent to any available solids surface. To what extent the dirt surface is made water-repellent could not yet be deduced from this result.

Table IV shows the adsorption of Eumulgin and ethylhexanol from a mixture of both components in a 1:1 ratio to coal and dirt. It can be seen that the adsorption of Eumulgin to dirt is not affected by the addition of ethylhexanol. The adsorption to coal, on the other hand, is considerably reinforced by ethylhexanol. In the reverse situation the adsorption of ethylhexanol to coal is much curtailed by Eumulgin, while the addition of Eumulgin has no effect on the adsorption to dirt.

Adsorbent	Adsorptive	Residual concentration C_R (mg/l)	Charging Q (mg/kg)
Gas coal (500 - 0 μm)	Eumulgin +(Ethylhexanol)	16.5	1 786
		7.7	878
		2.6	322
		1.3	187
Dirt (500 - 0 μm)	Eumulgin +(Ethylhexanol)	14.4	1 900
		4.1	948
		1.7	380
		0.3	203
Gas coal (500 - 0 μm)	Ethylhexanol +(Eumulgin)	49.6	1 003
		17.5	695
		1.8	354
		0.3	186
Dirt (500 - 0 μm)	Ethylhexanol +(Eumulgin)	91.5	175
		44.6	158
		16.7	75
		8.0	44

TABLE IV: Mixture adsorption isotherms of Eumulgin + ethylhexanol to coal and dirt

The results show that a displacement adsorption takes place here in favour of Eumulgin to coal. What repercussions this phenomenon has on flotation and in what way the non-ionogenous surfactant can take on the functions of flotation alcohol will have to be examined in further investigations. It might be said in anticipation that Eumulgin makes dirt particles enter flotation concentrates in larger measure, as the first tests showed. This points to Eumulgin acting as water-repellent on dirt.

If a polymer flocculant is added as another active component it is found that adsorption of ethylhexanol, previously greatly reduced by Eumulgin, is again improved in the presence of a polymer. Table V contains the measured values of this test series.

Adsorbent	Adsorptive	Residual concentration C_R (mg/l)	Charging Q (mg/kg)
Gas coal (125-315 μm)	Ethylhexanol +(Eumulgin)	81.5	366
		40.4	237
		12.5	140
		4.6	101
Gas coal (125-315 μm)	Ethylhexanol +(Eumulgin) +10 mg/l P 6000	80.2	413
		36.9	280
		12.1	147
		4.5	101

TABLE V: Mixed adsorption of ethylhexanol and Eumulgin in the presence of flocculant P 6000 to coal in the 125-315 m size range

The measurement results have shown, when taken together with the measured wetting temperatures, that polymer molecules interact with surfactant and flotation agent molecules and for their part act as adsorbents for surfactant and flotation agents. As with the adsorption isotherms the surfactants concentration of the solution is reduced so steeply that a positive effect can be seen with regard to alcohol adsorption. An adsorption of the flotation agent to polymers does have a negative effect for the flotation because polymer molecules do not only form unselective agglomerates from coal and dirt particles but also float down the concentration of the flotation agent in the suspension.

4. EFFECT OF FLOCCULANTS ON FLOTATION BEHAVIOUR

In view of the physical and chemical forces which have an important influence in the flotation process laboratory investigations were carried out to limit the surface-changing effect by additives. According to the 6-stage flotation analysis (10) there is a flotation kinetic progression for each concentrate stage, characterised by yield volume, ash content and size distribution. The yield and ash values with their differing time scales in the flotation process are used to derive the selectibility coefficient η (11) as outlined by Kubitza. With the aid of the selectivity coefficient η it is possible to describe flotation as an ongoing event at any point in time, i.e. the relation between the yield obtained and the theoretically possible yield is expressed by η, the residual amount remaining in the flotation cell at each concentrate stage and the yield amount being taken into account.

Figure 1 shows this interrelation using as an example a gas coal with 2-ethylhexanol-(1) as a flotation reagent and stepped-down flocculant concentrations from 0-20 mg/l.

Here it can be clearly seen how not only the separation efficiency but also flotation kinetics is reduced with increasing flocculant content. The results of these tests, which were carried out under the same conditions and with the originally named flotation agents and additives, have shown that with all the flotation agents and flocculants examined - be they of differing intensity - selectivity and partly flotation kinetics are progressively influenced for the worse as flocculant concentrations rise. These results point to polymer flocculants binding in coal and dirt particles unselectively on the one hand, but also acting as adsorbent for the flotation reagents and so reduce the effectiveness of these reagents, something which is confirmed by the results in measuring wetting temperatures and adsorption isotherms.

The interaction between flotation and sedimentation reagents was also noted during the flocculation process with dirt. Here it was found that the average setting speed declined and residue became thickened as flotation agent concentration increased. Coal was not completely sedimented under the influence of flotation reagents, while a part of the coal substance floated up in the sedimentation tank and formed a slurry layer that is difficult to breakup on the surface of the water.

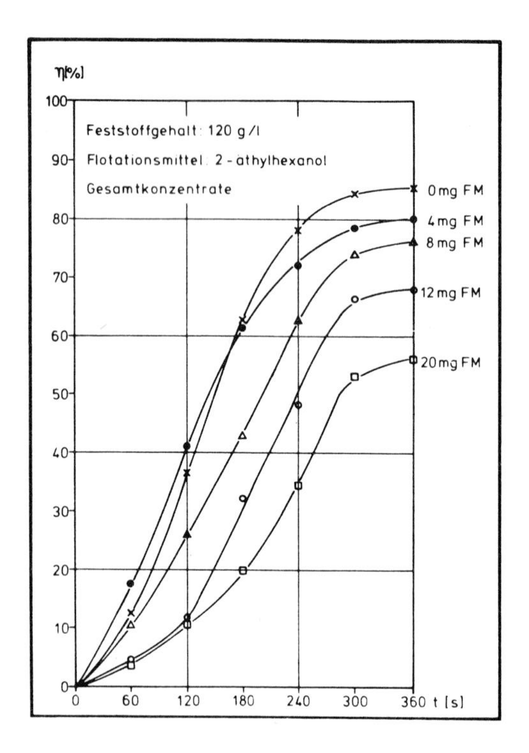

solids content: 120 g/l
flotation agent: 2-ethylhexanol
total concentrates

Figure 1: Selectivity-time dependence of the size range 500-0 μm with different flocculant contents

BIBLIOGRAPHY

1. E. Töpfer, U. Bilsing and H.J. Schulze: Freiberger Forschungshefte A 571 (1977), p. 67/77.

2. R.M. Horsley and H.G. Smith: Fuel 30, 3 (1951), p. 54/63.

3. G.A. Brady and A.W. Ganger: Industr. Eng. Chem. 32 (1940), p. 1599.

4. P.D. Luisenko: Coke and Chemistry (USSR) 5, 40 (1935); Chem. Abstr. 30, (1936), 6922.

5. H.A.J. Pieters: Brennst. Chem. 12 (1931) p. 325.

6. E. Bierbrauer, E.- and J. Pöpperle U.K. Patent No. 450, 044 (1935).

7. E.C. Plante and K.L. Sutherland — Amer. Inst. Min. (metall.) Energs. Techn. Publ. No. 2297 (1948).

8. H.F. Yancey and J.A. Taylor — U.S. Bur. Min. Rep. Invest. No. 3263 (1935).

9. E. Bierbrauer and J. Pöpperle — Glückauf 70 (1934), p. 933.

10. Flotation Analysis — DIN-Draft No. 22 017.

11. K.-H. Kubitza — Glückauf 116 (1980) No. 10, p. 508.

Afternoon discussion (by J.K. WILKINSON)

Mr CAMMACK (National Coal Board) remarked that there was an increasing requirement for classifying fine coal. He asked whether the work that Mr. DANIEL had described had yet found any commercial application and, if so, whether there were any problems with blinding of the screen mesh during continuous operation. Mr. DANIEL said that the technique was used in installations for sizing sand and kaolin. It had not yet been applied to coal, but discussions were being held with manufacturers in this connection.

Dr. HOBERG (Institut für Aufbereitung, Kokerei und Brikettierung) observed that the rotating probability screen had undergone a long period of development and had now been used by the National Coal Board for some years. He asked whether any studies had been made of the crushing effect that the machine might have, or of wear and tear on the screen. Dr. ARMSTRONG replied that development had been completed in 1979. The National Coal Board had 18 units and 12 had been installed overseas (in the USA, Australia and China) all of which, with one exception, were used to screen coal. Degradation tests had been carried out on the pilot scale and surprisingly, in view of the high speed of rotation of the screen, no size reduction had been observed. The explanation for this was thought to lie in the fact that the material treated was mostly below 30 mm and had a stable size consist. Degradation of material with a top size of 100 mm might be expected, but no tests had been conducted at that size level.

With reference to Mr. ERDMANN's paper, Dr. SCHULTZE (Gewerkschaft Auguste Victoria) stated that four vacuum belt filters had been installed at the Auguste Victoria mine about a year ago. Like the installation at Lohberg, they had a filter surface of $60m^2$ exposed to vacuum, but they were about 8m shorter because the belt width had been increased from 2.4 m to 3.2 m. For ash contents of 7.5 to 8% and a size distribution of 50 to 55% below 0.063 mm, residual water contents of 20 to 23% were achieved. The specific throughput was about 375 kg/m^2h. The installation was operated with very little or, if possible, no addition of flocculant. The belt speed and the feed rate were controlled as a function of the vacuum level selected. Mr. DANIEL (CERCHAR) asked Mr. ERDMANN if he had tried injecting water into the filter press and whether he considered that drum or disc filters gave better results in terms of residual moisture. Mr. ERDMANN said that no studies of water injection had been carried out. He thought that drum filters were to be preferred on the basis of better dewatering. They also gave improved solid output and could be operated continuously.

Mr. MUELLER (Saarbergwerke AG) considered that although Dr. HEY had demonstrated how flotation was influenced by the use of increased amounts of additives, the increases he had described were outside the range that could be used in operational practice. Dr. HEY explained that the first aim of the tests he had described had been to characterize interface behaviour. It had been shown that additions of 2mg/l or less gave improved flotation, but with lower selectivity, and that in terms of particle size even the smallest amounts of additive had a clear influence.

The following written question to Dr. HEY was submitted by Mr. BROOKES (University of Nottingham):

"I understand from work within the UK that the presence of anionic flocculants in flotation a) increases flotation induction time and b) increases yield with flocculant addition, but only up to a maximum, after which the yield falls dramatically. I would like the author's comments on these two points".

Dr. HEY referred Mr. BROOKES to the following publications:

Untersuchungen über die Möglichkeit zur Wiederverwendung von Abwässern aus die Kohleaufbereitung
M. Dočkal
Verfahrenstechnik 2J (1958)No.7, 283-287.

Influence of polyacrylamide on coal flotation
V.A. Ostriyi and D.I. Stepanova
Koks i Khimiya 1975, No.8, 3 -7.

Influence of polyoxethylene on coal-slurry flotation and filtration
A.A. Baichenko, N. Ya. Tarasova and A.A. Baichenko
Koks i Khimiya 1976, No.9, 10 - 13.

Wechselwirkungen von Flotations- und Sedimentationsreagenzien und ihre Bedeutung für die flotative Steinkohlensortierung
W. Hey, K.-H. Kubitza und D. Leininger
Glückauf-Forschungkefte 44 (1983) H.5, 226-233.

At the end of the discussion, Mr. LEMKE (Steinkohlenbergbauverein) made the following remarks at the invitation of the Chairman:

I must begin by thanking Mr. DE GREEF of the Directorate-General for Energy in the Commission of the European Communities for inviting me to this Conference. I did not expect to be invited as a matter of course, as I have been living now for more than ten years in so-called retirement, just over the other bank of the Sauer, a river which connects my country with this enchanting Grand Duchy. I appreciate very much being given the floor by the Chairman of the second technical session, and I - who am as it were a recent fossil from the early history of our European Coal and Steel Community - shall try to repay his courtesy by looking back to the start of the partnership of coal preparation engineers, which together we experienced and, to a certain extent, shaped.

Since there are few people left who experienced that time, and since the spirit which prevailed then can relax human relations generally, both now and in the immediate future, that partnership should not be forgotten, but extolled as a model.

Before Jean Monnet, who was born in Cognac in 1888, in the first half of this month, became involved in shaping the United States of Europe, the French had instituted the International Coal Preparation Congress, which met for the first time in Paris in 1950. They showed the Belgians, the Germans, the British, the Dutch and the Americans too, what could be done. We Germans counted it an honour, which we shall not forget, that they should at that Conference use our language at that time, as well as French and English - and so easily at that. It gives me great joy and satisfaction to recall that time before such an assembly: it is something I have long wanted to do.

As if it were yesterday, I can see before me the distinguished, dignified figure of Etienne Audibert, who commanded such respect. He it was who was ultimately responsible for the new approach and the new beginning. I can hardly resist a smile, when I recall how in the dark he used to roll a cigarette in his pocket with one hand and then, while he was talking, move it as if by magic from one corner of his mouth to the other without using his hands. Similarly, Cheradame, Belugou, Terra and others are names that will not be forgotten by my generation.

Why do I dwell so insistently on this event from that high summer of 1950 in Paris (where our shoes would often stick to the hot asphalt) ? Because it is a model of how a Community - of whatever kind - can be formed. It was marked by a spirit of mutual understanding which has not, and must not, be extinguished: the same spirit which characterizes and inspires our Conference here today.

There are two conditions, really, which have to be met at all times and in all places. First, differences in the participants' technical knowledge must be reduced by exchanging experience, as we are doing today, and each of the speakers' main arguments fully explored in discussion. This is a relatively easy task. The second essential, which is more difficult, is that the number of fully worked man-hours per head per year should be approximately the same in all parts of the Community one is trying to build.

In my view, we coal preparation engineers managed to achieve this - without really being fully aware of our success - by dint of tough, wholesome competition. By meeting both requirements, it became possible to span all frontiers and even link the five continents. We learnt to respect and help each other, and regard the members of other nations as friends.

In these dangerous times, may such a spirit succeed throughout the World. We must hope, and believe, and do everything to see, that it does.

Closing address

J.K. Wilkinson

Ladies and Gentlemen,

We have now reached the end of the short symposium on coal preparation. I hope everyone who has participated will have found it as interesting and useful as I have.

I should like to take the opportunity of these last few moments to convey my own thanks and those of the Commission to everyone who has helped to organize this meeting and to make it a success. I am particularly grateful to the chairman and the speakers, to the ECSC Coal Preparation Committee, which has also acted as the planning committee for the symposium, to the Luxembourg departments of the Commission's Directorate-General for Information Market and Innovation and Directorate-General for Personnel and Administration and the staff and interpreters of the Joint Service Interpretation-Conferences, and especially to Mr. Linster, Miss Goebel and Mr. Rotondo. Last but not least I should also like to thank Mrs. Swartenbroeckx who works with us in the Coal Directorate's Division for Research and Technology. The work of the Commission's services is not over yet, because they will be engaged in preparing the proceedings of the symposium, which will be sent to you all in due course.

It is my belief that the coal research carried out with ECSC aid and reported at meetings like the present one not only furthers the cause of science and of the coal industry, but also helps to bring the Community's member countries a little closer together. I hope that, in addition to learning something useful at the technical level, everyone has had the opportunity to renew personal and professional contacts and also to establish new ones, perhaps in a broader sphere. There is still an opportunity to do that at the reception which follows the meeting, and to which you are all cordially invited.

I now declare the symposium closed and wish everyone a safe return home after the reception, which will take place downstairs in the restaurant area, and which begins at 5.30

LIST OF PARTICIPANTS

ARMSTRONG, M.P.
Coal Preparation (R & D) Engineer
National Coal Board - MRDE
31, Lawn Avenue
GB - ETWALL, DERBY DE6 6JB

BACHMANN, C.
Diplomphysiker
Bergbau-Forschung GmbH
Franz-Fischer-Weg 61
D - 4300 ESSEN 13

BARTELT, D.
General Manager
Montan-Consulting GmbH
Division Surface Installations
Rüttenscheider Strasse 1
D - 4300 ESSEN 1

BERGER, F.
Betriebsingenieur
Gewerkschaft Sophia Jacoba
Steinkohlenbergwerk/Postfach
D - 51042 HUECKELHOVEN

BOCK, B.
Professor
Mechanische Verfahrenstechnik
Universität - GHS - Essen
Postfach 103764
D - 4300 ESSEN 1

BOUSSANGE, J.-P.
Ingénieur, HBCM
9, avenue Benoît Charvet/B.P. 67
F - 42002 ST. ETIENNE CEDEX

BREUER, H.
Prokurist
Maschinenfabrik Augsburg-Nürnberg AG
Unternehmensbereich GHH-STERKRADE
Dorstener Str. 250
D - 4630 BOCHUM

BROOKES, G.
Lecturer
Mining Department,
Nottingham University
Mining Engineering Department
University Park
GB - NOTTINGHAM NG7 2RD

BROWAEYS, P.
Ingénieur
INIEX-Pâturages
74, Chemin à Baraques
B - 7000 MONS

BRUNELLO, J.-M.
Ingénieur
CERCHAR
B.P.2
F - 60550 VERNEUIL EN HALATTE

BUCHET
N.V. Kempense Steenkolenmijnen
Grote Baan 27
B - 3530 HOUTHALEN

CAMBIER, F.
Université Catholique de Louvain
Faculté des Sciences Appliquées
Département thermodynamique
et turbomachines
2, Place du Levant
B - 1348 LOUVAIN-LA-NEUVE

CAMMACK, P.
M.R.D.E. National Coal Board
10, Vicarage Close, Winshill
GB - BURTON-ON-TRENT, Staffs. DE15 OBE

CUSTERS
N.V. Kempense Steenkolenmijnen
Grote Baan 27
B - 3530 HOUTHALEN

DANIEL, J.-L.
Ingénieur
CERCHAR
B.P.2
F - 60550 VERNEUIL EN HALATTE

DEFLANDRE
N.V. Kempense Steenkolenmijnen
Grote Baan 27
B - 3530 HOUTHALEN

DELOBELLE, J.
Secrétaire Général
Cerchar
33, rue de la Baume
F - 75008 PARIS

DE PIETRI
Engineer
Ansaldo S.p.A.
Via G. D'Annunzio 113
I - 16121 GENOVA

DERYCKE, E.
Chercheur
Université Libre de Bruxelles
Service Exploitation des Mines
C.P. 165
Avenue Roosevelt 50
B - 1050 BRUXELLES

DE WINTER
N.V. Kempense Steenkolenmijnen
Grote Baan 27
B - 3530 HOUTHALEN

d'HUART, B.
Représentant
Thomas Locker S.A.
17, rue de la Station
B - 1350 LIMAL-WAVRE

EDELMANN, G.
Diplomchemiker
Hoechst AG
Abt. Marketing TH-ATA
Postfach 800320
D - 6230 FRANKFURT/MAIN 80

ENGIN, K.
Diplomingenieur
PLE Kernforschungsanlage Jülich
Postfach 1913
D - 5170 JUELICH

ERDMANN, W.
Wissentschaftlicher Mitarbeiter
Bergbau-Forschung GmbH
Franz-Fischer-Weg 61
D - 4300 ESSEN

FAUTH, G.
Diplomphysiker
Bergbau-Forschung GmbH
Franz-Fischer-Weg 61
Postfach 130140
D - 4300 ESSEN 13

FORTUIN, B.
Project-manager
Neom B.V.
Postbus 17
NL - 6130 AA SITTARD

FUCHS, A.
Maschinendirektor
Gewerkschaft Sophia Jacoba
Steinkohlenbergwerk
Postfach
D - 5142 HUECKELHOVEN

FUNDA, P.
Diplomingenieur
Institut für Aufbereitung,
Kokerei und Brikettierung
Wüllnerstrasse 2
D - 5100 AACHEN

FUSENIG, V.
2. Vorsitzender
Jugend-Presse-Rat e.V.
Turmstr. 21
D - 5500 TRIER

GEURTS, A.
Chemicus
Ankersmit Holding B.V.
Postbus 260
NL - 6200 AG MAASTRICHT

GOERTZ, M.
Kaufmann
Hein, Lehmann AG
Fichtenstrasse 75
D - 4000 DUESSELDORF

GROESSEL, D.
Dr.-Ingenieur
Stahlwerke Peine-Salzgitter AG
Postfach 41 1180
D - 3320 SALZGITTER 41

GUETTINGER, M.
Geschäftsführer
Siebtechnik GmbH
Platanenallee 46
D - 4330 MUELHEIM-RUHR

HAMZA, H.A.
Manager, Coal Research Laboratory
Energy Mines & Resources, CANMET
P.O.Bag 1280
CANADA - DEVON, Alberta TOC 1E0

HASSE, W.
Diplomingenieur
KHD Humboldt Wedag AG
Herner Str. 299
D - 463 BOCHUM

HERZOG, W.
Diplomchemiker
Staatlisches Materialprüfungsamt
Marsbruchstr. 186
D - 4600 DORTMUND 41

HESSE, A.
Manager
Century Oils (Deutschland)
Justus-Von-Liebig-Str. 48
D - 6057 DIETZENBACH

HEY, W.
Diplomchemiker
Bergbauforschung Essen GmbH
Franz-Fischer-Weg 61
D - 4300 ESSEN

HEYLEN, P.
Project Engineer
CEN-SCK
Boeretang 200
B - 2400 MOL

HOBERG, P.
Dr.-Ingenieur
Institut für Aufbereitung,
Kokerei und Brikettierung
Wüllnerstrasse 2
D - 5100 AACHEN

HOLLAND, T.
Director
Nationalised Sector
Century Oils Ltd.
P.O.Box 2
Century St.
Hanley
GB - STOKE-ON-TRENT ST1 5HU

HUBLET, J.
Administrateur-Directeur
S.A. Birtley N.V.
Avenue de Tervueren 32-36
Btes 11-12
B - 1040 BRUXELLES

HULSBOSCH
N.V. Kempense Steenkolenmijnen
Grote Baan 27
B - 3530 HOUTHALEN

JACKERS
N.V. Steenkolenmijnen
Grote Baan 27
B - 3530 HOUTHALEN

JAMAR, A.
Directeur Commercial
Thomas Locker S.A.
17, rue de la Station
B - 1350 LIMAL-WAVRE

JANITSCHEK, R.
Diplomingenieur
Verfahrenstechnik
Bergbau AG Niederrhein
Abt. T3.4
Baumstrasse 31
D - 4100 DUISBURG 17

JENKINSON, D.E.
Chief Coal Preparation Engineer
National Coal Board
Mining Research & Development Est.
Ashby Road, Stanhope Bretby
GB - BURTON-ON-TRENT DE15 OQD

JONES, G.
Chartered Mechanical Engineer
National Coal Board
Mining Research & Development
Establishment
Ashby Road, Stanhope Bretby
GB - BURTON-ON-TRENT DE15 OQD

JUNGMANN, A.
Ingenieur
M.A.N.
Maschinenfabrik Augsburg-Nürnberg AG
Unternehmensbereich GHH-Sterkrade
Dorstener Str. 250
D - 4630 BOCHUM

KAH, M.
Ingénieur
Houillères du Bassin de Lorraine
2, rue de Metz
F - 57802 FREYMING-MERLEBACH

KAMPF, A.
Bergoberrat
Oberbergamt für das Saarland und
das Land Rheinland-Pfalz
Am Staden 17
D - 6600 SAARBRUECKEN

KERKDIJK, C.B.W.
Physicist
FDO
P.O.Box 379
NL - 1000 AJ AMSTERDAM

KOCH, K.
Diplomingenieur
Kernforschungsanlage Jülich/PLE 2
Postfach 1913
D - 5170 JUELICH

KROENER, H.
Betriebsdirektor
Preussag AG Kohle
Postfach 1464
D - 4530 IBBENBUEREN

LAMMERTINK, E.
Tech. Manager
Dow Chemical B.V.
P.O.Box 48
NL - 4530 AA TERNEUZEN

LEININGER, D.
Wissenschaftlicher Mitarbeiter
Bergbau-Forschung GmbH
Franz-Fischer-Weg 61
D - 4300 ESSEN

LEMKE, K.
Dr.-Ingenieur
Steinkohlenbergbauverein
Schloss-Strasse 16
D - 5521 BOLLENDORF

LEONHARD, J.
Diplomingenieur
Bergbau-Forschung-Essen
Abt. Aufbereitung
Franz-Fischer-Weg
D - 4300 ESSEN

LIMPACH, R.
Chef de service adjoint
ARBED - Recherches
Case postale 141
66, rue de Luxembourg
L - ESCH-SUR-ALZETTE

LINDEN, J.
Société P. Worth
Rue d'Alsace
L - LUXEMBOURG

LOTZ, A.
Abteilungsleiter
Maschinenfabrik Augsburg-Nürnberg AG
Unternehmensbereich GHH Sterkrade
Dorstener Strasse 250
D - 4630 BOCHUM

LUEDKE, H.
Wissenschaftlicher Mitarbeiter
Diplom-Ingenieur
Bergbau-Forschung GmbH
Franz-Fischer-Weg 61
D - 4300 ESSEN 13

MAJCHROWICZ,B.
Research Scientist
Limburgs Universitair Centrum
Universitaire Campus
B - 3610 DIEPENBEEK

MANACKERMAN, M.
Area Coal Preparation Engineer
North Derbys. Area
National Coal Board
GB - BOLSOVER NEAR CHESTERFIELD S44 6AA

MAUFORT
N.V. Kempense Steenkolenmijnen
Grote Baan 27
B - 3530 HOUTHALEN

MILES, N.
Lecturer
Nottingham University
Mining Engineering Department
University Park
GB - NOTTINGHAM NG7 2RD

MUELLER, D.
Diplomingenieur
Saarbergwerke AG
Trierer Str. 1
D - 6600 SAARBRUECKEN

NANZ, H.
Diplomingenieur
Krupp Polysius AG
Postfach 2340
D - 4720 BECKUM

NATER, K.A.
Director
SBN
P.O.Box 151
NL - 6470 ED EYGELSHOVEN

NEUHAUS, H.
Dr. Ingenieur
RAG
Feldmannstrasse 53
D - 6600 SAARBRUECKEN

NEUMANN, U.
Leiter der Technischen Abteilung
Eschweiler Berwerks-Verein AG
Roermonderstrasse
D - 5120 HERZOGENRATH

POL, F.
Ingénieur
Usinor Dunkerque
59, rue du Comte Jean
F - GRANDE SYNTHE

PADBERG, W.
Diplomingenieur
Saarbergwerke AG
Trierer Str. 1
D - 6600 SAARBRUECKEN

PIGNON, B.
Ingénieur
Houillères du Bassin de Lorraine
2, avenue Emile Huchet
F - 57800 FREYMING MERLEBACH

PUT, J.
N.V. Steenkolenmijnen
Grote Baan 27
B - 3530 HOUTHALEN

RAUCHS, A.
Ingénieur
ARBED
79, rue de Luxembourg
L - 4221 ESCH-SUR-ALZETTE

REGGERS, G.
Research Scientist
Limburgs Universitair Centrum
Universitaire Campus
B - 3610 DIEPENBEEK

REICHERT, A.
Institut für Aufbereitung,
Kokerei und Brikettierung
Wüllnerstrasse 2
D - 5100 AACHEN

REINEKING, R.
Dr.-Ingenieur
Ruhrkohle AG Essen
Drosselweg 17
D - 4250 BOTTROP

REINHARD, A.
Ingénieur
Houillères du Bassin de Lorraine
2, avenue Emile Huchet
F - 57800 FREYMING MERLEBACH

REUTER, J.M.
Diplomingenieur
Chemische Fabrik Stockhausen GmbH
Postfach 570
D - 4150 KREFELD

ROESNER, W.
Diplombergingenieur
Exploration und Bergbau GmbH
Abteilung Kohle
Steinstr. 20
D - 4000 DUESSELDORF 1

RZONZEF, L.
Ingénieur
Administration des Mines
84, avenue de Am Cortil
B - 4040 ESNEUX

SCHACHT, N.
Diplomingenieur
PLE Kernforschungsanlage Jülich
Postfach 1913
D - 5170 JUELICH

SCHMIDT, P.
Professor
Universität Essen, FB 13
D - 4300 ESSEN 1

SCHNEIDER, D.
Diplomingenieur
Institut für Aufbereitung,
Kokerei und Brikettierung
Wüllnerstrasse 2
D - 5100 AACHEN

SCHNITTGER, B.
Fahrsteiger
Preussag AG Kohle
Postfach 1464
D - 4530 IBBENBUEREN

SCHULTZE, H.-H.
Dr.-Ingenieur, Oberingenieur
Gewerkschaft Auguste Victoria
Postfach 1180
D - 4370 MARL

SEIFERT, G.
Diplomingenieur
Deutsche Nalco Chemie GmbH
Hamburger Allee 2-10
D - 6000 FRANKFURT 90

SMIRNOW, S.
Professor Dr.-Ingenieur
Westfälische Berggewerkschaftskasse
Herner Str. 45
D - 4630 BOCHUM

STAHL, H.
Diplomchemiker
Erz- und Kohleflotation GmbH
Herner Str. 299
Postfach 102729
D - 4630 BOCHUM 1

STEPHENSON, K.
Managing Director
Century Oils (Europe) S.A.
Heideveld 54
B - 1511 BEERSEL- HUIZENGEN

STRAGIER
N.V. Kempense Steenkolenmijnen
Grote Baan 27
B - 3530 HOUTHALEN

SUPP, A.
Dr. Ingenieur
Krupp Polysius AG
Postfach 2340
D - 4720 BECKUM

THIELEN, G.
Leitender Bergdirektor
Oberbergamt für das Saarland
und das Land Rheinland-Pfalz
Am Staden 17
D - 6600 SAARBRUECKEN

THOMAS, F.
Ingénieur
Chef Services Techniques
S.A. Retraitement des Terrils
du Centre
120, Chaussée Brunehault
B - 7120 PERONNES

TOWNSEND, F.
Chief Engineer
Century Oils Ltd.
P.O.Box 2
Century St., Hanley
GB - STOKE-ON-TRENT ST1 5HU

TSANG, I.
Engineer
Babcock Power Ltd.
Research Centre
High Street
GB - RENFREW PA4 8UW (Scotland)

ULVELING, L.
P. Wurth Compagnie
Rue d'Alsace
L - LUXEMBOURG

UNGER, G.
Maschinen- und Apparatenbau v. Humboldt
Heidestrasse
D - 5628 HEILIGENHAUS

VAN POUCKE, L.
Professor
Limburgs Universitair Centrum
Universitaire Campus
B - 3610 DIEPENBEEK

VAN SWIJGENHOVEN, H.
Research Scientist
Limburgs Universitair Centrum
Universitaire Campus
B - 3610 DIEPENBEEK

VON ACHTEN, R.
Diplomingenieur, Board Member
Braunschweigische Maschinenbauanstalt AG
Am Alten Bahnhof 5
D - 3300 BRAUNSCHWEIG

WADE, W.G.
Coal Preparation Engineer
Davy McKee (Stockton) Ltd.
Ashmore House
GB - STOCKTON-ON-TEES, Clevel.TS18 3RE

WAUTERS, P.
Université Catholique de Louvain
Faculté des Sciences Appliquées
Département thermodynamique et turbomachines
2, Place du Levant
B - 1348 LOUVAIN-LA-NEUVE

WILCZYNSKI, P.
Dr.-Ingenieur
Bergbau AG Lippe/T 3.4 Aufbereitung
Virchowstr. 129
D - 4650 GELSENKIRCHEN

WILKINSON, J.K.
Commission des Communautés européennes
DG XVII/B3
200, rue de la Loi
B - 1049 BRUXELLES

WINGERTSCHES, H.
Ingenieur
Deutsche Nalco Chemie GmbH
Hamburger Allee 2-10
D - 6000 FRANKFURT 90

ANDERSONIAN LIBRARY
WITHDRAWN
FROM
LIBRARY
STOCK
UNIVERSITY OF STRATHCLYDE

2